Entrepreneurial Innovations, Models, and Implementation Strategies for Industry 4.0

This book explores the link between entrepreneurship and innovation, providing an understanding of the latest developments in the field. It explores numerous challenges to entrepreneurship, such as failures and socio-economic issues, and presents concepts, models, and implementation strategies for Industry 4.0.

Industry 4.0 presents unique challenges and unique opportunities to entrepreneurs. Social, techno, and women's entrepreneurship, among other business sectors and domains, are discussed in this new context. This book highlights the strategies of successful start-up organizations, focusing on the most critical factors for starting a new enterprise in today's business environment, and provides efficient remedies for common problems. These strategies and solutions will help build a sustainable development model for developing economies and contribute to a global awareness of the entrepreneurship environment.

Entrepreneurial Innovations, Models, and Implementation Strategies for Industry 4.0 is both a practical guide for new entrepreneurs to establishing start-ups and an enlightening collection of research for graduate students, postgraduate students, scholars, and academicians across all disciplines. Policymakers will also benefit from this book's understanding of entrepreneurial ecosystems and their applications.

Information Technology, Management and Operations Research Practices

Series Editors:
Vijender Kumar Solanki, Sandhya Makkar, and Shivani Agarwal

This new book series encompasses theoretical and applied books and is aimed at researchers, doctoral students, and industry practitioners to help solve real-world problems. The books will help in the various paradigms of management and operations and discuss the concepts and emerging trends on society and businesses. The focus is to collate the recent advances in the field and take the readers on a journey that begins with understanding the buzz words like employee engagement, employer branding, mathematics, operations, technology, and how they can be applied in various aspects. It walks readers through engaging with policy formulation, business management, and sustainable development through technological advances and provides a comprehensive discussion on the challenges, limitations, and solutions of everyday problems such as how to use operations, management, and technology to understand the value-based education system, health and global warming, and real-time business challenges. The book series brings together some of the top experts in the field throughout the world who contribute their knowledge regarding different formulations and models. The aim is to provide the concepts of related technologies and novel findings to an audience that incorporates specialists, researchers, graduate students, designers, experts, and engineers who are occupied with research in technology, operations, and management-related issues.

Sustainability, Big Data, and Corporate Social Responsibility
Evidence from the Tourism Industry
Edited by Mohammed El Amine Abdelli, Nadia Mansour, Atilla Akbaba, and Enric Serradell-Lopez

Entrepreneurial Innovations, Models, and Implementation Strategies for Industry 4.0
Edited by Ravindra Sharma, Geeta Rana, and Shivani Agarwal

Multi-Criteria Decision Models in Software Reliability
Methods and Applications
Edited by Ashish Mishra, Nguyen Thi Dieu Linh, Manish Bhardwaj, and Carla M. A. Pinto

For more information about this series, please visit: https://www.routledge.com/Information-Technology-Management-and-Operations-Research-Practices/book-series/CRCITMORP

Entrepreneurial Innovations, Models, and Implementation Strategies for Industry 4.0

Edited by

Ravindra Sharma, Geeta Rana, and Shivani Agarwal

CRC Press
Taylor & Francis Group
Boca Raton London New York

CRC Press is an imprint of the
Taylor & Francis Group, an **informa** business

First edition published 2023
by CRC Press
6000 Broken Sound Parkway NW, Suite 300, Boca Raton, FL 33487-2742

and by CRC Press
4 Park Square, Milton Park, Abingdon, Oxon, OX14 4RN

CRC Press is an imprint of Taylor & Francis Group, LLC

Library of Congress Cataloging-in-Publication Data
Names: Sharma, Ravindra (Professor of management), editor. | Rana, Geeta,
editor. | Agarwal, Shivani, editor.
Title: Entrepreneurial innovations, models, and implementation strategies
for Industry 4.0 / edited by Ravindra Sharma, Geeta Rana, and Shivani Agarwal.
Description: First edition. | Boca Raton : CRC Press, 2022. |
Includes bibliographical references and index. |
Summary: "The book will include discussions of social, women, and techno entrepreneurship along with introducing strategies adopted for Industry 4.0. Insights on entrepreneurship strategies and challenges are presented as well as how to develop and create models for different business sectors and domains. The book also provides a resource to select effective and efficient remedies for entrepreneurial problems which can result in a sustainable development model for developing economies and also add to a global platform of awareness and knowledge in the global entrepreneurship environment"— Provided by publisher.
Identifiers: LCCN 2022010254 (print) | LCCN 2022010255 (ebook) |
ISBN 9781032107936 (hardback) | ISBN 9781032107967 (paperback) |
ISBN 9781003217084 (ebook)
Subjects: LCSH: Technological innovations—Economic aspects. |
Industry 4.0. | Entrepreneurship.
Classification: LCC HC79.T4 E595 2022 (print) | LCC HC79.T4 (ebook) |
DDC 338/.064—dc23/eng/20220718
LC record available at https://lccn.loc.gov/2022010254
LC ebook record available at https://lccn.loc.gov/2022010255

ISBN: 978-1-032-10793-6 (hbk)
ISBN: 978-1-032-10796-7 (pbk)
ISBN: 978-1-003-21708-4 (ebk)

DOI: 10.1201/9781003217084

Typeset in Times
by codeMantra

Contents

Preface

Throughout the world, a panic scenario is being developed where most of us have become helpless in finding the panacea for our anomalies. Along with health and safety, the corona virus has imposed a threat to our survival due to heavy losses in terms of jobs and various development activities. The significance of finding better prospects becomes the need of the hour, but moreover, it also reflects the uproar that has been created due to the epidemic.

In the direction to attain life sustainability, entrepreneurship can act as a pivotal tool in finding every possible way to change this current emergency into a blessing in disguise. The Fourth Industrial Revolution, also known as Industry 4.0, has provided an unprecedented platform for entrepreneurial innovation in various spheres. Industry 4.0 is the name of the automation and data exchange trends in factory technology. This term includes the physical–cyber systems, the Internet, and the cloud and cognitive computing. Industry 4.0 affects business models and implementation strategies of businesses and brings a new horizon to entrepreneurial development.

Now the definition of entrepreneurship is not limited to starting a business anymore; it has revolutionized the spirit to find better avenues where we can challenge and come up with effective remedies for all our odds.

Entrepreneurship can redefine the model of development in such a way that its efficacy cannot be ignored. It can create great avenues so that people can put themselves into the mainstream and likewise be competitive enough to secure their survival in the long run.

Entrepreneurship has always paved the way for potential businesspeople to have the courage to come out of their nutshells and get their notions to be more dynamically tested where they can stand against all odds.

This book on entrepreneurial innovations, models, and implementation strategies for Industry 4.0 focuses on determining various elements and investigating all possibilities to provide a sustainable platform for aspiring entrepreneurs. It also entails educating students on a variety of entrepreneurship-related issues, such as failures, planning, strategy, socioeconomic, technical, and financial support. The aim of this book is to expand the knowledge of the latest entrepreneurial activities and their importance in multiple aspects.

This book is an anthology of 10 research papers that provide a rich repertoire of tools and techniques across business functions researched, tested, and validated in various business settings. The chapters highlight technology disruption in entrepreneurship, entrepreneurial competencies, digital economy, techno-entrepreneurship, green entrepreneurship, women entrepreneurship venture capital financing, entrepreneurship challenges of the manufacturing sector in Industry 4.0, and exploring various cases. It also provides an overview of the most popular and frequently used models and techniques in entrepreneurial development. Overall, this book brings forth a new stream of thoughts by a few fine researchers in the domain of entrepreneurship. We wish and hope that this book will generate interest among not only budding entrepreneurs, but also academicians and researchers.

Happy Reading!
Editors,

Dr. Ravindra Sharma
Dr. Geeta Rana
Dr. Shivani Agarwal

Acknowledgments

We are thankful to the Almighty for the blessings to complete this book successfully.

The book in its present shape is possible due to the keen interest shown by academic colleagues from around the world. We would like to express our gratitude and appreciate the contributions of the authors.

A large number of our friends and colleagues from academia and industry offered their support to make this project successful. This support has come in many forms: review and re-review of chapters, suggestions for revisions and modifications, and critical comments to make the chapters more thorough and meaningful. We acknowledge their tremendous support and are thankful for their valuable comments.

We acknowledge all the support, inspiration, and motivation of our faculty colleagues and remain indebted for their outstanding intellectual efforts. Our gratitude is due to Dr. Vijay Dhasmana, the Vice-Chancellor and Dr. Alok Saklani, Dean, Himalayan School of Management Studies, both of Swami Rama Himalayan University, Jolly Grant Dehradun, India, for the faith, trust and responsibility empowered in us and for extending their full support and being a constant source of inspiration and pillars of strength throughout this pursuit.

Our special thanks to Dr. Santosh Rangaker, Indian Institute of Technology, Roorkee, India and Dr. Jeevan Jyoti, University of Jammu, India for their constant encouragement and support in this endeavor. We thank our prospective readers in advance for they would be a source of improvement and further development of this book.

We express our gratitude to our publisher CRC (Taylor & Francis USA) and the entire editorial team who lent their wonderful support throughout and ensured timely processing of the manuscript and publishing the book.

Finally, we would fail in our duty if we did not acknowledge the most loving care and cooperation enjoyed by us, all through the working on this book, from our dear families for selfless support provided while compiling this book.

Editors,

Dr. Ravindra Sharma
Dr. Geeta Rana
Dr. Shivani Agarwal

Editors

Ravindra Sharma is an Assistant Professor in the Himalayan School of Management Studies at Swami Rama Himalayan University, Dehradun, India. He has earned his Ph.D. in Management from Uttarakhand Technical University, India. He holds Master's degrees in business administration (MBA) and computer applications (MCA) and was also qualified for the University Grant Commission National Eligibility Test in Management (UGC-NET). Dr. Sharma has more than 16 years of corporate and academic experience. He has conducted many faculty development programs for national and international audiences. He has published research papers in refereed journals of Emerald, Sage, Springer, IGI Global, and Inderscience and has published books in the areas of employer branding, HR 4.0, Internet of Things (IoT), and artificial intelligence (AI) with reputed publishers such as Taylor & Francis Group USA, Nova Science Publishers USA, and IRP Publication House India. Dr. Sharma has contributed chapters to books published by Springer IGI Global, Taylor & Frances, and Palgrave Macmillan. He has presented research papers at national and international conferences and has also attended a paper development workshop at the Indian Institute of Management Rohtak, India. He has been honored as a session chair and keynote speaker in international conferences at various universities. His research interests include employer branding, entrepreneurship, Industry 4.0, and talent management.

Geeta Rana is an Associate Professor at Himalayan School of Management Studies in Swami Rama Himalayan University, Jolly Grant, Dehradun, India. She earned her Ph.D. in management from the Indian Institute of Technology (IIT, Roorkee). She also completed a certification course in HR Analytics from the Indian Institute of Management Rohtak (IIMR). She is engaged in teaching, research, and consultancy assignments and has more than 17 years of experience in teaching and handling various administrative as well as academic positions. She has to her credit over 50 papers published in refereed journals of Emerald, Sage, Springer, Taylor & Francis, Elsevier, and Inderscience publishers and has also presented several research papers at national and international conferences. Dr. Rana has contributed several chapters to books published by Cambridge UK, Springer IGI Global, and Palgrave Macmillan. She has also authored a book titled *Counseling Skills for Managers* published by New Age Publications. She has conducted and attended various workshops, faculty development programs (FDPs) and management development programs (MDPs). Dr. Rana has received many awards. Her research interests include data analytics, Human Resource Management 4.0, knowledge management, managerial effectiveness, justice, values, employer branding, and innovation.

Shivani Agarwal is an Assistant Professor in KIET School of Management at KIET Group of Institutions, Ghaziabad, India. She earned her Ph.D. from the Indian Institute of Technology (IIT, Roorkee) in Management. She is engaged in teaching, research, and consultancy assignments. She has more than 10 years of experience in teaching and handling various administrative as well as academic positions. She also presented several research papers at national and international conferences. Dr. Agarwal has contributed chapters to books published by Taylor & Francis, Springer, and IGI Global. She has conducted and attended workshops, FDPs and MDPs. She is the Book Series Editor of Information Technology, Management & Operations Research Practices, CRC Press, Taylor & Francis Group, USA. She is a guest editor with IGI Global, USA. Her research interests include quality of work-life, trust, subjective well-being, knowledge management, employer branding innovation, and human resource management.

Contributors

Shivani Agarwal
KIET School of Management
KIET Group of Institutions
Ghaziabad, India

Widad Ahl Maatallah
Tangier International Higher Institute of Tourism (ISITT)
Tangier, Morocco

Afaf Akhter
Department of Business Administration
Aligarh Muslim University
Aligarh, India

Kashif Ali
Universiti Teknologi PETRONAS (UTP)
Malaysia

Zaiba Ali
College of Business and Administration
Princess Nourah Bint Abdulrahman University
Riyadh, Saudi Arabia

Suha Bilquis
Department of Business Administration
Aligarh Muslim University
Aligarh, India

Mostapha El Idrissi
Cadi Ayyad University
Marrakesh, Morocco

Ayesha Farooq
Department of Business Administration
Aligarh Muslim University
Aligarh, India

Vipul Jain
School of Management & Commerce Studies
Shri Guru Ram Rai University
Dehradun, India

Satirenjit Kaur Johl
Universiti Teknologi PETRONAS (UTP)
Malaysia

Saurabh Joshi
Himalayan School of Management Studies
Swami Rama Himalayan University, Dehradun, India

Pavnesh Kumar
Department of Management Sciences
Mahatma Gandhi Central University
Bihar, India

S. Senthil Kumar
Department of Organizational Behavior/ Human Resource Management
Institute of Management Technology
Nagpur, India

Esther Mwiti
School of Business and Management Studies
Technical University of Kenya (TUK)
Nairobi, Kenya

Bhakti Parashar
Vellore Institute of Technology-Bhopal University
Bhopal, India

Pushpendu Rakshit
Amity Business School Mumbai
Amity University
Panvel, India

Geeta Rana
Himalayan School of Management Studies
Swami Rama Himalayan University
Dehradun, India

Roomi Rani
Govt. SPMR College of Commerce
Jammu, India

Ashwini Kumar Saini
G. B. Pant Institute of Engineering and Technology
Pauri, India

Antima Sharma
University of Engineering & Management
Jaipur, India

Jeevesh Sharma
Department of Business Administration
Manipal University
Jaipur, India

Ravindra Sharma
Himalayan School of Management Studies
Swami Rama Himalayan University
Dehradun, India

Chandan Veer
Mahatma Gandhi Central University
Bihar, India

1 Entrepreneurial Competencies Needed to Improve Innovation Performance in Industry 4.0

An Artificial Neural Network Approach

Kashif Ali and Satirenjit Kaur Johl
Universiti Teknologi PETRONAS (UTP)

CONTENTS

DOI: 10.1201/9781003217084-1

1.1 INTRODUCTION

Innovation promoted by the Industry 4.0 (I4.0) continues to transform the business landscape that requires a skillful workforce capable of adopting new technological trends. This digital transformation process drastically reduces the employment share of the low-skilled workforce (Kruger and Steyn, 2020). On the other hand, the incorporation of digital technologies into business operations increases the demand for a digital-skilled workforce, which requires the digital upskilling and reskilling of the workforce (Ali & Johl, 2022). In addition to this, the COVID-19 pandemic has further amplified employees' needs for digital skills and competencies (APFC, 2020). As a result, employees' abilities must be adapted to new production methods with a more emphasis on interdisciplinary competencies. Furthermore, employees are capable of solving and addressing the problems linked with I4.0 (Chaka, 2020). The concept of I4.0 poses a significant challenge for businesses in terms of the application of I4.0 technologies and human interaction with these technologies (Ali & Johl, 2021).

The concept of I4.0 is described as the digitalization of industrial and manufacturing processes accomplished through the use of cutting-edge technology and tools (Kerin and Pham, 2019). I4.0 is characterized by a network of various digital devices that collect and disseminate information between process and business units to achieve higher performance (Ali & Johl, 2021). Likewise, automated systems and machines improve business operations and processes (Lytras et al., 2017). The ultimate focus of I4.0 is to improve manufacturing efficiency, productivity, and quality (Shet and Pereira, 2021). Different business sectors like manufacturing, logistics, and supply chain are prominent in adopting and implementing I4.0 technologies. As a result, considerable research is being conducted in both human-focused and technological domains. However, to remain successful and effective in the era of digitalization, there is a pressing need to identify and prioritize human competencies, especially from the entrepreneurial perspective (Kruger and Steyn, 2020).

It is generally argued that an industrial revolution (IR) occurs when new ways of understanding and technologies emerge, promoting revolutionary changes both in economy and in society (Schwab, 2017). The first IR was started in England in around 1780 and was characterized by the use of the steam engine. The second IR was initiated around 1870 by scientific breakthroughs such as the use of steel, electricity, radio, telephones, and mass production (Souza et al., 2021). Around 1969, the third IR was started and was generally known as the digital revolution because computers and the internet emerged, which transformed the global landscape. In comparison to the first three IRs, the I4.0 is more complex as it is characterized by a variety of technologies that have an influence on speed and connectivity (Philbeck and Davis, 2018). On the basis of the foregoing debate, it is argued that I4.0 is driven by algorithm-based decision-making and demands sophisticated employees' competencies in order to be successful. This is in contrast to the previous three, which developed at their own speed.

From the human capital perspective, the acute shortage of trained individuals in the manufacturing industry represents a problem for intelligence-based industrial operations. A recent survey by Udemy (2020) acknowledged that 83% of employees worldwide believe that new and innovative technologies demand new digital skills. According to the report of Society for Human Resource Management (2019), the US economy is confronted with a widening skill gap, which undermines the country's long-term economic prosperity. Furthermore, around 75% of organizations have difficulty in finding skillful employees during the recruitment process. The most frequently missing skills are problem-solving and creativity (37%), the ability to deal with complexity (32%), and communication (31%). Similarly, QS Global skills gap report (2019) affirmed that problem-solving, teamwork, and communication skills are most important. Additionally, Asia–Pacific Economic Cooperation (2020) stated that the current pandemic situation has disrupted the labor market because the existing workforce lacks digital skills. As a result, there is an urgent need to invest in developing the workforce's digital capabilities. The following paragraph highlights some underlying issues that are necessary in the current context.

According to Ivanov et al. (2020), I4.0 concept is multidisciplinary because many concepts come from different disciplines like computer sciences, engineering, and business and management. It is argued that these areas are fully interconnected both at operational and business levels. Therefore, to succeed in this complex environment, business owners and employees need to adopt new ways of thinking. Additionally, to implement digitalization within organization, workforce will need to continuously update their knowledge, skills, competencies, etc. In this context, enterprises would use the existing competency framework to identify the skills and competencies needed in I4.0. Furthermore, the organizational workforce has certain skills, but they are unable to prioritize them. Hence, this research study identifies the entrepreneurial competencies that are necessary in the I4.0 context from existing framework and prioritizes them to further enhance the innovation performance.

In the prior literature, I4.0 technologies have received much attention; however, the entrepreneurial competencies needed for success in this new ecosystem have received little attention. Consequently, there is a lack of knowledge of the entrepreneurial skills necessary to be successful in digitalization. Although academics have noted a dearth of debate on functional competencies, the importance of these skills and capabilities from an entrepreneurial viewpoint remains ignored in a holistic manner. Therefore, this study addressed the following research questions in the light of these research gaps.

RQ1. What are the entrepreneurial key competencies within the I4.0 perspective?

RQ2. What is the level of importance of these digital competencies with regard to innovation performance?

Therefore, this research was conducted to identify the key entrepreneurial competencies for I4.0 and to prioritize these digital competencies with innovation performance. To illustrate this, the article is organized into various sections. The next section presents a brief review of the literature. The methods and analysis are presented in the next part. The discussion of research findings is presented, followed by study conclusion.

1.2 LITERATURE REVIEW

This section introduces the ideas of entrepreneurial competencies, I4.0, and innovation performance in order to clarify the subjects covered in this study.

1.2.1 Entrepreneurial Competencies and I4.0

In the strategic management literature, the term 'competency' has been extensively used. It is defined as the knowledge, abilities, and attitudes needed to perform a specific task (Baum et al., 2001). Volery et al. (2013) argued that an entrepreneur can learn and attain the competencies through training, learning, and practices. Thus, entrepreneurial competences are described as "the fundamental characteristics, such as personality traits, motivations, specific knowledge, social roles, skills, and self-images that are essential for successful entrepreneurship" (Bird, 1995; Mitchelmore and Rowley, 2010). These fundamental characteristics have been recognized as the entrepreneurial competencies (Osman and Rahim, 2014) or individual traits that enable them to carry out activities in the most beneficial ways (Lazar and Paul, 2015). An entrepreneur is someone who starts a business and adds value to it by identifying opportunities and utilizing resources (Tehseen et al., 2019). Furthermore, the entrepreneurial competencies are multidimensional in that they help entrepreneurs adopt I4.0 technologies in their operations (Okolie et al., 2021; Hecklau et al., 2017; Prifti et al., 2017).

There is a paradigm shift in I4.0 in terms of organizational structures and individuals' roles and activities. As a result of this shift, the importance of human elements must be recognized, as they are the most important component of digitalization (Flores et al., 2020). Furthermore, it also changes the workforce interactions that impact individual activities, as they are needed to become more coordinated, creative, and strategic (Kazancoglu and Ozkan-Ozen, 2018).

The review study of Flores et al. (2020) highlights the main competence of humans for I4.0: education/training, competencies, wellbeing, and innovation. Gehrke et al. (2015) presented the expectation from I4.0 employees in three tires. The first tire refers to the tools and technologies that the skilled workforce utilizes. The second tire refers to tasks that I4.0 employees need to perform. The third tire is the core of the model and refers to skills and qualifications that are needed for I4.0 workforce. The scoping review of Chaka (2020) categorized the 21st-century skills into soft (communication, creativity, and problem-solving) and hard skills (programming and data analysis). Kurger and Steyn (2020) conceptualized the entrepreneurial competencies needed to effectively utilize novel I4.0 technologies. These competencies are innovation, creativity, leadership and communication, networking and sales, and integrate business and technology skills. According to Kipper et al. (2021), the competencies required for I4.0 include creativity, problem-solving, communication, innovation, leadership, teamwork, adaptability, flexibility, and self-management.

1.2.2 Entrepreneurship, I4.0, and Innovation

In today's increasingly complex and fast-changing environment, the strategic management literature recognizes that innovation has a significant role in organizations to create value and maintain competitive advantage. Organizations with a higher level of innovation will be more responsive to changing environment and develop new skills that will enable them to perform better. In previous studies, employee knowledge, expertise, and devotion have been frequently cited as critical inputs in the value generation process and innovation initiatives.

The association between entrepreneurship and innovation has been extensively researched and well-documented in the literature. However, the majority of the literature is concerned with the ability of the entrepreneur to innovate (Mazzei, 2018; Schmitz et al., 2016). As a result, the entrepreneurship in emerging economies helps to improve economic conditions as well as create new jobs (Mamabolo et al., 2017). In this regard, the association between entrepreneurship and innovation is essential for economic development in I4.0 (Ibarra et al., 2018). In order to deal with competition and survival in the digital society, an innovative process must be developed and implemented (Lu et al., 2017). In I4.0, entrepreneurs can look at innovation from two different angles. The first viewpoint describes innovation as evolutionary process that refers to incremental and continuous advancement in technology or processes with the aim to increase efficiency. The second viewpoint describes innovation as a revolutionary process that is completely new and a disruptive technological and process development that can lead to additional area of revenue (Lu et al., 2017; Maree and McKenzie, 2015).

Both entrepreneurship and innovation concepts can address the rapid advancement in I4.0 technologies and can be used as a means to boost economic growth. Kruger and Steyn (2020) affirmed the association between entrepreneurship competencies and innovation in three logical layers of I4.0 technologies. The first layer of I4.0 technologies is the physical layer. This layer explains how entrepreneur competencies are associated with hardware technologies including sensors to detect physical changes. The second layer explains how entrepreneurs are skillful in operating connectivity devices such as Bluetooth, IP networks, radio-frequency identifiers (RFIDs), or Wi-Fi. The final layer explains how employees are competent in analyzing and processing the data that comes from the physical layers through connective devices. The technologies included in the final layer are artificial intelligence, cloud systems, and big data.

1.3 MATERIAL AND METHODS

This chapter describes two parts of the methodological procedures, building a research portfolio and a survey study, as shown in Figure 1.1.

1.3.1 Research Portfolio Construction

Based on the study objective of key entrepreneurial competencies in I4.0, this study conducted a systematic literature review based on Pagani et al. (2015) and Campos et al. (2018), as well as Souza et al. (2021). Based on the application of the Ordination Index (InOrdination) equation, which organizes the portfolio using three criteria: impact factor (IF), number of citations (Ci), and year of publication, this permits the formation of portfolios of scientifically relevant publications. Pagani et al. (2015) proposed nine steps of the Methodi Ordination as described in Table 1.1.

1.3.2 Quantitative Study

In order to identify the key entrepreneurial competencies in the I4.0, first, this study performed a thorough literature review. The identified competencies (Table 1.2) were further analyzed through survey research.

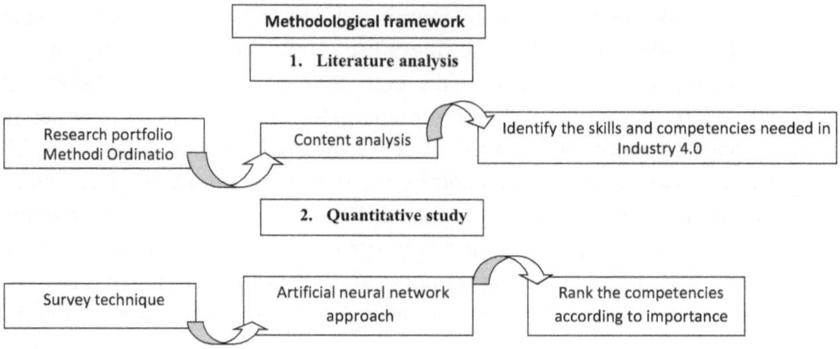

FIGURE 1.1 Methodological framework.

TABLE 1.1
Nine Steps of the Method Ordination

Steps	Description	Details
1	Defining the research purpose	The goal of the research was to locate publications that addressed a specific issue relating to the theme of entrepreneurial competencies and I4.0.
2	Performing preliminary database search	Scopus and Google scholar
3	Study keywords and database selection	**Keywords**: ("Industry 4.0" OR "fourth industrial revolution" OR 4IR OR "Industrie 4.0") AND (Skills OR Competencies) **Databases**: Scopus, Web of Science, IEEE Xplore, and Google Scholar
4	Final search in database	Article year; 2011–2020 (2011 as inaugural year for I4.0); and article published in English language only. Total articles in portfolio = 21
5	Procedures for filtering (document types, duplicate, unrelated papers to the theme)	Deletion of duplicate articles = 42 Articles that are not related to the theme = 47 Final articles in the portfolio = 21
6	Determine the impact factor, publication year, and number of citations for each article in the portfolio	**IF** (Journal citation reports (JCR), Scopus cite score) **Number of citations** (WOS, Scopus, Google Scholar)
7	Use the InOrdination equation to categories the publication InOrdination $= \left(\begin{array}{l} \text{IF}/1,000 + \alpha \times \\ \left[10 - (\text{Research year} - \text{Publish year})\right] + \sum Ci \end{array} \right)$	**Alfa value (α)** = The researcher assigned a score from 1 to 10 based on the importance of the topic's novelty. The study focuses on competencies in I4.0, the criterion year becomes important to define the value 10 for alfa. Articles with an In Ordination higher and equal to 80 were selected.

(Continued)

TABLE 1.1 (*Continued*)
Nine Steps of the Method Ordination

Steps	Description	Details
8	Final and download the complete articles	The full version of all 21 articles specified in the previous stage was found. Appendix
9	Final reading and articles systematic analysis	The full portfolio of order articles has been obtained and content reading, and methodical analysis have begun. The identification of key competencies for I4.0 is shown in Table 1.2.

Source: Author's own.

TABLE 1.2
Key Competencies for I4.0

Sr. #	Skills	References
1	Technical skills (TSs)	Benešová and Tupa (2017), Collet et al. (2015), Hecklau et al. (2016), Gray et al. (2018), Saniuk et al. (2021), Jerman et al. (2020), Graczyj-Kucharska et al. (2018), Liboni et al. (2019), Kannan & Garad (2020)
2	Problem-solving (PS)	Chaka (2020), Benešová and Tupa (2017), Caruso (2018), Gocmen & Coskun (2019), Demartini and Benussi (2017), Eguchi (2014), Hecklau et al. (2016), Krpalek and Krelova (2016), Mohamed (2018), Saniuk et al. (2021), Khan et al. (2021), Grzybowska and Łupicka (2017), Kazancoglu and Ozkan-Ozen (2018), Kipper et al. (2021), Prifti et al. (2017), Shet and Pereira (2021), Flores et al. (2020), Meyer et al. (2015), Hecklau et al. (2017)
3	Communication skills (CSs)	Chaka (2020), Caruso (2018), Collet et al. (2015), Demartini and Benussi (2017), Eguchi (2014), Hecklau et al. (2016), Krpalek and Krelova (2016), Magruk (2016), Schiuma (2017), Slusarczyk (2018), Tvenge et al. (2016), Saniuk et al. (2021), Khan et al. (2021), Kipper et al. (2021), Kruger and Steyn (2020), Graczyj-Kucharska et al. (2018), Prifti et al. (2017), Liboni et al. (2019), Flores et al. (2020), Meyer et al. (2015), Hecklau et al. (2017)
4	Creativity (CR)	Chaka (2020), Benešová and Tupa (2017), Caruso (2018), Eguchi (2014), Hecklau et al. (2016), Kortuem et al. (2013), Krpalek and Krelova (2016), Schiuma (2017), Jerman et al. (2020), Khan et al. (2021), Grzybowska and Łupicka (2017), Kipper et al. (2021), Kruger and Steyn (2020), Graczyj-Kucharska et al. (2018), Prifti et al. (2017), Hecklau et al. (2017)

(*Continued*)

TABLE 1.2 (*Continued*)
Key Competencies for I4.0

Sr. #	Skills	References
5	Technological/digital innovation (TI)	Caruso (2018), Collet et al. (2015), Mohamed (2018), Morrar et al. (2017), Slusarczyk (2018), Umachandran et al. (2018), Jerman et al. (2020), Kipper et al. (2021), Kruger and Steyn (2020)
6	Big data analysis and programming skills (BDA)	Chaka (2020), Benešová and Tupa (2017), Gocmen & Coskun (2019), Magruk (2016), Kipper et al. (2021), Graczyj-Kucharska et al. (2018), Prifti et al. (2017), Shet and Pereira (2021), Hecklau et al. (2017)

Source: Author's own.

1.3.2.1 Data Collection

Based on the research objectives, an online survey technique was used to collect data from Malaysian small and medium (SMEs) manufacturing firms operating in different states. The Federation of Malaysian Manufacturers (FMM) list was used to draw the study sample. The FMM publishes a list of about 3,300 small and medium-sized manufacturing and service enterprises (Lim et al., 2021). The data was collected using the stratified simple random sampling technique. Based on the states, the sample is divided into strata. As a result, the selected sample can be regarded a valid representative of the entire population. The questionnaires were distributed to the respondent that holds the position of senior management and above. Because they have hands-on experience with entrepreneurial skills and innovative performance. Prior to distribution, the information was validated by four industry practitioners and two professors in the subject of entrepreneurship. A pilot study was done with 30 respondents to examine the clarity of the scale items and the overall design of the questionnaire. A total of 125 completed questionnaires were received, reflecting a response rate of 20.3%. However, 10 questionnaires have missing and incomplete information, leaving 115 viable questionnaires. The study required at least 109 respondents with the G*power application and as 115 questionnaires were received, the sample size is more than acceptable in this situation.

1.3.2.2 Operationalization of Measurement Items

A five-point Likert scale was used to score the evaluation items, with the exception of questions about demographic information and the respondents' companies. The four items of technical skills were adopted from Smith et al. (2007) and Buntat et al. (2016). The problem-solving construct was measured using four items adopted from Jabri (1991) and Wang (2013). The five items of communication skills and three items of creativity were adopted from Abdul (2018). The four items of technological/digital innovation were adopted from Perdomo-Ortiz et al. (2007) and Sá and Abrunhosa (2007). The big data analysis and programming construct were measured using five items adopted from Mikalef et al. (2019). On the other hand, the dependent construct (innovation performance) was measured from four items adopted from Chen and Huang (2009).

1.4 ANALYSIS AND RESULTS

1.4.1 Demographic Analysis

Table 1.3 contains a comprehensive demographic breakdown of the respondents. According to the gender breakdown, there were more males (65.22%) than females (34.78%). In terms of age distribution, 23.48% of respondents were under the age of 30, 29.56% were between the ages of 31 and 40, 38.26% were between the ages of 41 and 50, and 8.70% were over the age of 51. Furthermore, 20% of respondents have an intermediate qualification, 40% have a graduate degree, 28.70% have a master's degree, and the remaining 11.30% have some other qualification. Finally, 32.17% had fewer than five employees, other enterprises had between 5 and <75 (40%), and some employed between 75 and 200 (27.83%).

1.4.2 ANN Analysis

The artificial neural network technique is considered the most intelligent of the available analytical techniques. The ANN demonstrated a large but complex network that contains multiple neurons distributed into three layers; input, hidden, and output layers (Hew et al., 2018). Models like multivariate regression analysis (MRA) and structural equation modeling (SEM) cannot represent the complexity of human decision-making since these analytical methods only find the linear

TABLE 1.3
Demographic Profile

Variable	Items	Frequency	Percentage (%)
Gender	Male	75	65.22
	Female	40	34.78
Age	20–30	27	23.48
	31–40	34	29.56
	41–50	44	38.26
	≥51	10	8.70
Education	Intermediate	23	20.00
	Graduation	46	40.00
	Master/doctorate	33	28.70
	Other	13	11.30
Number of employees in organization	Less than 5	37	32.17
	5–<75	46	40.00
	75–≤200	32	27.83

Source: Author's own analysis.

relation (Liébana et al., 2018; Chong & Chang, 2012; Sim et al., 2014; Tan et al., 2014; Liébana-Cabanillas et al., 2017). Additionally, MRA and SME are compensating models assuming that decline in one variable may be compensated by an addition of another variable (Wong et al., 2020). In the research, the independent/exogenous variables are not compensable. This means that a drop in one skill or competency cannot be substituted by an increase in another because all constructs are distinct in terms of conceptualization and definitions; therefore, these constructs are not identical. In addition, ANNs are inappropriate for testing and assessing causal relationships between exogenous and endogenous variables due to their "black-box" nature (Tan et al., 2014; Chong & Chan, 2012; Hew et al., 2016). Similar to Lim et al. (2021), the application of ANN in this research is utilized to evaluate each predictor variable relative importance. This technique outperforms linear models in terms of multicollinearity, homoscedasticity, and non-normality of distribution (Hew et al., 2018; Wong et al., 2020). ANN models have actually surpassed traditional statistical techniques like MRA and SEM due to their high degree of prediction accuracy (Leong et al., 2015).

Prior research has attempted to provide a more detailed description of ANNs. Haykin (2001) stated that ANN is a massively parallel distributed processor composed of simple units with a neural propensity for accumulating and storing experimental knowledge and making available for use. In a later study, Haykin (2004) stated that ANN analysis was similar to the human brain performing a particular function or tasks. This technique is employed in a variety of research fields like supply chain quality management (Lim et al., 2021), Blockchain in SME operations (Wong et al., 2020), m-commerce (Liébana-Cabanillas et al., 2017), social media addiction (Leong et al., 2019), and e-learning (Sharma et al., 2017). However, its applicability to entrepreneurship in order to create organizational success is limited. Therefore, using ANN analysis to the predictive power of exogenous constructs to explain the endogenous construct, this work intends to make a significant methodological contribution.

An ANN model's architecture is made up of three layers; input, hidden, and output. The root mean square errors (RMSEs) and normalized significance of the input neurons were determined using the feed-forward-back-propagation technique and multilayer perceptrons (Hew et al., 2018). To address model fit, similar to Leong et al. (2020) and Lim et al. (2021), the researcher assigned 90% of the data for training and 10% for testing. This study used a 10-fold cross-validation process to avoid the possibility of overfitting and obtained the RMSE values (Ooi & Tan, 2016). The average RMSE values for the training and testing methods are 0.133 and 0.084, respectively, as shown in Table 1.4.

As a result of the sensitivity analysis, Table 1.5 was used to evaluate each predictor variable according to its relative importance. According to the findings of the sensitivity analysis, creativity has the greatest impact on innovation performance, with a 100% normalized relative importance, followed by CS (48%), PS (46%), BDA (25%), TS (24%), and TI (21%).

TABLE 1.4
RMSE Values for Neural Network

Training		Testing		Total Samples
N	RMSE	N	RMSE	
99	0.147	16	0.116	115
107	0.151	8	0.056	115
104	0.155	11	0.115	115
99	0.107	16	0.045	115
107	0.148	8	0.130	115
102	0.126	13	0.062	115
102	0.132	13	0.076	115
104	0.099	11	0.086	115
102	0.129	13	0.077	115
102	0.135	13	0.077	115
Mean	0.133		0.084	
S.D	0.019		0.028	

Source: Author's own analysis.

TABLE 1.5
Sensitivity Analysis

Neural Network	TS	BDA	CS	CR	TI	PS
NN-1	0.123	0.102	0.544	1.000	0.102	0.434
NN-2	0.104	0.134	0.315	1.000	0.098	0.324
NN-3	0.258	0.098	0.137	1.000	0.324	0.377
NN-4	0.391	0.132	0.125	1.000	0.093	0.251
NN-5	0.034	0.009	0.744	1.000	0.043	0.164
NN-6	0.231	0.321	0.411	1.000	0.319	0.560
NN-7	0.474	0.342	0.159	0.051	0.230	1.000
NN-8	0.338	0.220	0.693	1.000	0.290	0.611
NN-9	0.079	0.184	1.000	1.000	0.243	0.145
NN-10	0.030	0.612	0.012	1.000	0.059	0.126
Mean importance	0.206	0.215	0.414	0.868	0.180	0.399
Normalized importance	24%	25%	48%	100%	21%	46%

Source: Author's own analysis.

1.5 DISCUSSION

Based on the systematic review of the literature and empirical study, this section discusses the findings in a more detailed way. Additionally, the discussion on the results is organized to answer the research questions. Underscoring these two RQs, the research has two purposes; identifying the skills and competencies in I4.0 by the 21 journal articles; and normalized importance of these competencies toward innovation performance. To address the first RQ; a research portfolio was constructed to identify the skills and competencies needed in I4.0 are technical, problem-solving, communication, creativity, technological/digital innovation, and big data analysis and programming. Then, to address the second RQ, the ANN technique was used to analyze the normalized importance of competencies toward innovation performance. Based on the results, creativity is the most important competency, followed by problem-solving, communication, big data analysis and programming, and technical and technological innovation. Based on the analysis, creativity, problem-solving, and communication are the top three competencies needed in I4.0. In the following section, these three are discussed in more detail.

1.5.1 Creativity

The term creativity is described as the individual ability to understand the business environment in new directions, uncover hidden information, and link and generate solution between the apparently unrelated phenomena. The ability to think creatively is more important in today's digital economy. This result is in line with studies such as Grzybowska and Łupicka (2017), Santos et al. (2021) and Fitsilis et al. (2018). According to World Economic Forum, by 2022, the importance of creativity skill is expected to rise (WEF, 2018). Similarly, the creativity skill remains important and falls in the top 10 skills of 2025 (WEF, 2020).

1.5.2 Problem-Solving

In the literature, problem-solving and complex problem-solving are interchangeably used. The problem-solving skill involves both the analytical and creative skills. In digital era, problem-solving is essential both at the workplace and in personal situations. This result is in line with studies such as Grzybowska and Łupicka (2017) and Chaka (2020). In the problem-solving skills, entrepreneurial apply their logic and intelligent abilities to solve the problem. Problem-solving skill is in great demand in a number of industries. The problem-solving skills remain highly essential in the I4.0 era and fall in the top 10 skills of 2022 (WEF, 2018) and in 2025 (WEF, 2020).

1.5.3 Communication

The entrepreneurship studies highlight the importance of communication skills (Hood and Young, 1993). Furthermore, communication skills are also important in I4.0. The findings of this study are in line with Hecklau et al. (2017) and Meyer et al. (2015). The traditional communication pyramid represents a typical hierarchy of interaction among customers and organizations. On the other hand, in I4.0, the communication pyramid transforms into heterogeneous social network (Zhang et al., 2015). The I4.0-enabled communication is able to achieve flexibility, accountability, and openness to knowledge and input of every member. Furthermore, it promotes a decentralized approach by creating an opportunity to contact and interact with all members (Flores et al., 2020).

1.6 CONCLUSIONS

I4.0 alters the way people and production systems collaborate with one another. The difference between I4.0 and preceding IR is the speed and interconnectedness of technologies. I4.0 technologies are changing the ways people work, socialize, and interact with one another. As a result, business owners have the option to digitalize and invent new initiatives. This study's goal is to identify and rank the most important entrepreneurial competencies in I4.0. Based on the empirical findings in I4.0, creativity, problem-solving, and communication are the top three entrepreneurial competencies.

REFERENCES

Ali, K., & Johl, S. K. (2022). Impact of Total Quality Management on SMEs Sustainable Performance in the Context of Industry 4.0. *Proceedings of International Conference on Emerging Technologies and Intelligent Systems*, 608–620. https://doi.org/10.1007/978-3-030-82616-1_50.

Ali, K., & Johl, S. K. (2021). Soft and hard TQM practices: future research agenda for industry 4.0. *Total Quality Management & Business Excellence*, 1–31. https://doi.org/10.1080/14783363.2021.1985448.

Abdul, O. E. (2018). Entrepreneurial skills and growth of Small and Medium Enterprise (SMEs): A comparative analysis of Nigerian entrepreneurs and Minority entrepreneurs in the UK. *International Journal of Academic Research in Business and Social Sciences*, *8*(5). https://doi.org/10.6007/ijarbss/v8-i5/4083.

Asia-Pacific Economic Cooperation. (2020). *APEC Closing the Digital Skills Gap Report: Trends and Insights*. APEC-Human Resources Development Working Group.

Baum, J. R., Locke, E. A., & Smith, K. G. (2001). A multidimensional model of venture growth. *The Academy of Management Journal*, *44*(2), 292–303. https://doi.org/10.2307/3069456.

Benešová, A., & Tupa, J. (2017). Requirements for Education and Qualification of People in Industry 4.0. *Procedia Manufacturing*, 11, 2195–2202. https://doi.org/10.1016/j.promfg.2017.07.366.

Bird, B. (1995). Toward a Theory of Entrepreneurial Competency. In J. A. Katz & R. H. Brockhaus, Sr. (Eds), *Advances in Entrepreneurship, Firm Emergence, and Growth*. 2, 51–72. Greenwich, CN: JAI Press.

Buntat, Y., Saud, M. S., Mokhtar, M., Kamin, Y., & Feh, L. S. (2016). Important skills for biomedical services: The perspectives of Malaysian employers and employees. *Work*, *55*(2), 481–487. https://doi.org/10.3233/wor-162417.

Caruso, L. (2018). Digital innovation and the fourth industrial revolution: epochal social changes? *AI & SOCIETY*, *33*(3), 379–392. https://doi.org/10.1007/s00146-017-0736-1.

Chen, C.-J., & Huang, J.-W. (2009). Strategic human resource practices and innovation performance — The mediating role of knowledge management capacity. *Journal of Business Research*, *62*(1), 104–114. https://doi.org/10.1016/j.jbusres.2007.11.016.

Chaka, C. (2020). Skills, competencies and literacies attributed to 4IR/industry 4.0: Scoping review. *IFLA Journal*, *46*(4), 369–399. https://doi.org/10.1177/0340035219896376.

Chong, A. & Chan, F. (2012). Structural equation modeling for multi-stage analysis on Radio Frequency Identification (RFID) diffusion in the health care industry. *Expert Systems with Applications*. *39*. 8645–8654. 10.1016/j.eswa.2012.01.201.

Collet, C., Hine, D., & du Plessis, K. (2015). Employability skills: perspectives from a knowledge-intensive industry. *Education & Training*, *57*(5), 532–559. https://doi.org/10.1108/et-07-2014-0076.

de Campos, E. A. R., Pagani, R. N., Resende, L. M., & Pontes, J. (2018). Construction and qualitative assessment of a bibliographic portfolio using the methodology Methodi Ordinatio. *Scientometrics*, *116*(2), 815–842. https://doi.org/10.1007/s11192-018-2798-3.

Fitsilis, P., Tsoutsa, P., & Gerogiannis, V. (2018). Industry 4.0: Required personnel competences. *Industry 4.0*, 3(3), 130–133.

Flores, E., Xu, X., & Lu, Y. (2020). Human capital 4.0: A workforce competence typology for industry 4.0. *Journal of Manufacturing Technology Management*, *31*(4), 687–703. https://doi.org/10.1108/jmtm-08-2019-0309.

Gehrke, L., Kühn, A., Rule, D., Moore, P., Bellmann, C., & Siemes, S. (2015). A discussion of qualifications and skills in the factory of the future: A German and American Perspective. *Hannover Messe 2015*. https://doi.org/10.1515/auto-2015-0068

Göçmen, Ö., & Coşkun, H. (2019). The effects of the six thinking hats and speed on creativity in brainstorming. *Thinking Skills and Creativity*, *31*, 284–295. https://doi.org/10.1016/j.tsc.2019.02.006.

Graczyk-Kucharska, M., Szafranski, M., Golinski, M., Spychala, M., & Borsekova, K. (2018). Model of competency management in the network of production enterprises in industry 4.0—assumptions. *Lecture Notes in Mechanical Engineering*, 195–204. https://doi.org/10.1007/978-3-319-68619-6_19.

Gray, J., Gerlitz, C., & Bounegru, L. (2018). Data infrastructure literacy. *Big Data & Society*. https://doi.org/10.1177/2053951718786316.

Grzybowska, K., & Łupicka, A. (2017). Key competencies for industry 4.0. *Economics and Management Innovations (ICEMI)*. Published. https://doi.org/10.26480/icemi.01.2017.250.253.

Haykin, S. (2001). *Neural Networks: A Comprehensive Foundation*. Englewood Cliffs, NJ: Prentice Hall.

Haykin, S. S. (2004), *Neural Networks: A Comprehensive Foundation* (2nd ed.). Englewood Cliffs, NJ: Prentice Hall.

Hecklau, F., Orth, R., Kidschun, F., & Kohl, H. (2017). Human resources management: Meta-study-analysis of future competences in industry 4.0. In *Proceedings of the International Conference on Intellectual Capital, Knowledge Management & Organizational Learning* (pp. 163–174).

Hew, T.-S. and Syed Abdul Kadir, S.L. (2016). Predicting instructional effectiveness of cloud-based virtual learning environment. *Industrial Management & Data Systems*, *116*(8), 1557–1584. https://doi.org/10.1108/IMDS-11-2015-0475.

Hew, J.-J., Leong, L.-Y., Tan, G. W.-H., Lee, V.-H., & Ooi, K.-B. (2018). Mobile social tourism shopping: A dual-stage analysis of a multi-mediation model. *Tourism Management*, 66, 121–139. https://doi.org/10.1016/j.tourman.2017.10.005.

Hood, J. N., & Young, J. E. (1993). Entrepreneurship's requisite areas of development: A survey of top executives in successful entrepreneurial firms. *Journal of Business Venturing*, 8, 115–135.

Ibarra, D., Ganzarain, J., & Igartua, J. I. (2018). Business model innovation through Industry 4.0: A review. *Procedia Manufacturing*, *22*, 4–10. https://doi.org/10.1016/j.promfg.2018.03.002.

Ivanov, D., Tang, C. S., Dolgui, A., Battini, D., & Das, A. (2020). Researchers' perspectives on Industry 4.0: multi-disciplinary analysis and opportunities for operations management. *International Journal of Production Research*, 1–24. https://doi.org/10.1080/00207543.2020.1798035.

Jabri, M. M. (1991). The development of conceptually independent subscales in the measurement of modes of problem solving. *Educational and Psychological Measurement*, *51*(4), 975–983. https://doi.org/10.1177/0013164491051004l7.

Jerman, A., Bertoncelj, A., Dominici, G., Pejić Bach, M., & Trnavčević, A. (2020). Conceptual key competency model for smart factories in production processes. *Organizacija*, *53*(1), 68–79. https://doi.org/10.2478/orga-2020-0005.

Kannan, K. S. P., & Garad, A. (2020). Competencies of quality professionals in the era of industry 4.0: A case study of electronics manufacturer from Malaysia. *International Journal of Quality & Reliability Management*, *38*(3), 839–871. https://doi.org/10.1108/ijqrm-04-2019-0124.

Kazancoglu, Y., & Ozkan-Ozen, Y. D. (2018). Analyzing workforce 4.0 in the fourth industrial revolution and proposing a road map from operations management perspective with fuzzy DEMATEL. *Journal of Enterprise Information Management*, *31*(6), 891–907. https://doi.org/10.1108/jeim-01-2017-0015.

Kerin, M., & Pham, D. T. (2019). A review of emerging industry 4.0 technologies in remanufacturing. *Journal of Cleaner Production*, 237, 117805. https://doi.org/10.1016/j.jclepro.2019.117805.

Kipper, L. M., Iepsen, S., dal Forno, A. J., Frozza, R., Furstenau, L., Agnes, J., & Cossul, D. (2021). Scientific mapping to identify competencies required by industry 4.0. *Technology in Society*, *64*, 101454. https://doi.org/10.1016/j.techsoc.2020.101454.

Kruger, S., & Steyn, A. A. (2020). A conceptual model of entrepreneurial competencies needed to utilise technologies of industry 4.0. *The International Journal of Entrepreneurship and Innovation*, *22*(1), 56–67. https://doi.org/10.1177/1465750320927359.

Lazar, N., & Paul, G. (2015). Entrepreneurial competencies in a business enterprise – An overview. *International Journal of Scientific Research*, *4*, 305–308.

Leong, L.-Y., Hew, T.-S., Lee, V.-H., & Ooi, K.-B. (2015). An SEM – artificial-neural-network analysis of the relationships between SERVPERF, customer satisfaction and loyalty among low-cost and full-service airline. *Expert Systems with Applications*, *42*(19), 6620–6634. https://doi.org/10.1016/j.eswa.2015.04.043.

Leong, L.-Y., Hew, T.-S., Ooi, K.-B., Lee, V.-H., & Hew, J.-J. (2019). A hybrid SEM-neural network analysis of social media addiction. *Expert Systems with Applications*, *133*, 296–316. https://doi.org/10.1016/j.eswa.2019.05.024.

Leong, L. Y., Hew, T. S., Ooi, K. B., & Wei, J. (2020). Predicting mobile wallet resistance: A two-staged structural equation modeling-artificial neural network approach. *International Journal of Information Management*, *51*, 102047. https://doi.org/10.1016/j.ijinfomgt.2019.102047.

Liboni, L. B., Cezarino, L. O., Jabbour, C. J. C., Oliveira, B. G., & Stefanelli, N. O. (2019). Smart industry and the pathways to HRM 4.0: implications for SCM. *Supply Chain Management: An International Journal*, *24*(1), 124–146. https://doi.org/10.1108/scm-03-2018-0150.

Liébana-Cabanillas, F., Marinkovic, V., Ramos de Luna, I., & Kalinic, Z. (2018). Predicting the determinants of mobile payment acceptance: A hybrid SEM-neural network approach. *Technological Forecasting and Social Change*, *129*, 117–130. https://doi.org/10.1016/j.techfore.2017.12.015.

Liébana-Cabanillas, F., Marinković, V., & Kalinić, Z. (2017). A SEM-neural network approach for predicting antecedents of m-commerce acceptance. *International Journal of Information Management*, *37*(2), 14–24. https://doi.org/10.1016/j.ijinfomgt.2016.10.008.

Lim, A. F., Lee, V. H., Foo, P. Y., Ooi, K. B., & Wei-Han Tan, G. (2021). Unfolding the impact of supply chain quality management practices on sustainability performance: An artificial neural network approach. *Supply Chain Management: An International Journal*, (ahead-of-print). https://doi.org/10.1108/scm-03-2021-0129.

Lu, J. G., Hafenbrack, A. C., Eastwick, P. W., Wang, D. J., Maddux, W. W., & Galinsky, A. D. (2017). "Going out" of the box: Close intercultural friendships and romantic relationships spark creativity, workplace innovation, and entrepreneurship. *Journal of Applied Psychology*, *102*(7), 1091–1108. https://doi.org/10.1037/apl0000212.

Lytras, M. D., Raghavan, V., & Damiani, E. (2017). Big Data and Data Analytics Research. *International Journal on Semantic Web and Information Systems*, *13*(1), 1–10. https://doi.org/10.4018/ijswis.2017010101.

Mamabolo, M. A., Kerrin, M., & Kele, T. (2017). Entrepreneurship management skills requirements in an emerging economy: A South African outlook. *The Southern African Journal of Entrepreneurship and Small Business Management*, *9*(1), 10. https://doi.org/10.4102/sajesbm.v9i1.111.

Maree, L., & McKenzie, K. (2015). Exploring the elements of strategic innovation drivers in South African banks. *Journal of Economic and Financial Sciences*, *8*(2), 604–620. https://doi.org/10.10520/EJC177422.

Mazzei, M. J. (2018). Strategic entrepreneurship: Content, process, context, and outcomes. *International Entrepreneurship and Management Journal*, *14*(3), 657–670. https://doi.org/10.1007/s11365-018-0512-9.

Meyer, G., Brünig, B., & Nyhuis, P. (2015). Employee competences in manufacturing companies – an expert survey. *Journal of Management Development*, *34*(8), 1004–1018. https://doi.org/10.1108/jmd-06-2014-0056.

Mikalef, P., Boura, M., Lekakos, G., & Krogstie, J. (2019). Big data analytics and firm performance: Findings from a mixed-method approach. *Journal of Business Research*, *98*, 261–276. https://doi.org/10.1016/j.jbusres.2019.01.044.

Mitchelmore, S., & Rowley, J. (2010). Entrepreneurial competencies: A literature review and development agenda. *International Journal of Entrepreneurial Behavior & Research*, *16*(2), 92–111. https://doi.org/10.1108/13552551011026995.

Okolie, U. C., Igwe, P. A., Ayoola, A. A., Nwosu, H. E., Kanu, C., & Mong, I. K. (2021). Entrepreneurial competencies of undergraduate students: The case of universities in Nigeria. *The International Journal of Management Education*, *19*(1), 100452. https://doi.org/10.1016/j.ijme.2021.100452

Ooi, K.-B., & Tan, G. W.-H. (2016). Mobile technology acceptance model: An investigation using mobile users to explore smartphone credit card. *Expert Systems with Applications*, *59*, 33–46. https://doi.org/10.1016/j.eswa.2016.04.015.

Osman, C. A., & Rahim, H. L. (2014). The relationship of entrepreneurial competencies and entrepreneurial intention of disabled youth: A study on behavioural and managerial competencies. *Australian Journal of Basic Applied Sciences*, *8*(23), 62–69.

Pagani, R. N., Kovaleski, J. L., & Resende, L. M. (2015). Methodi Ordinatio: a proposed methodology to select and rank relevant scientific papers encompassing the impact factor, number of citation, and year of publication. *Scientometrics*, *105*(3), 2109–2135. https://doi.org/10.1007/s11192-015-1744-x.

Patrícia Moura E Sá & Ana Abrunhosa (2007). The role of TQM practices in technological innovation: The Portuguese footwear industry case, *Total Quality Management & Business Excellence*, *18*(1–2), 57–66. DOI: 10.1080/14783360601043179.

Perdomo-Ortiz, J., González-Benito, J., & Galende, J. (2009). The intervening effect of business innovation capability on the relationship between Total Quality Management and technological innovation. *International Journal of Production Research*, *47*(18), 5087–5107. https://doi.org/10.1080/00207540802070934.

Philbeck, T., & Davis, N. (2018). The fourth industrial revolution: Shaping a new era. *Journal of International Affairs*, *72*(1), 17–22. https://www.jstor.org/stable/26588339.

Prifti, L., Knigge, M., Kienegger, H., & Krcmar, H. (2017). *A Competency Model for "Industrie 4.0" Employees.*

QS. (2019). *2019 Global Skills Gap Report.* Quacquarelli Symonds Ltd.

Saniuk, S., Caganova, D., & Saniuk, A. (2021). Knowledge and skills of industrial employees and managerial staff for the industry 4.0 implementation. *Mobile Networks and Applications.* Published. https://doi.org/10.1007/s11036-021-01788-4.

Santos, G., Sá, J. C., Félix, M. J., Barreto, L., Carvalho, F., Doiro, M., Zgodavová, K., & Stefanović, M. (2021). New needed quality management skills for quality managers 4.0. *Sustainability*, *13*(11), 6149. https://doi.org/10.3390/su13116149.

Schmitz, A., Urbano, D., Dandolini, G. A., de Souza, J. A., & Guerrero, M. (2016). Innovation and entrepreneurship in the academic setting: a systematic literature review. *International Entrepreneurship and Management Journal*, *13*(2), 369–395. https://doi.org/10.1007/s11365-016-0401-z.

Schwab, K. (2017). The Fourth Industrial Revolution. Switzerland: Crown. ISBN-13: 978-1-944835-01-9

Sharma, S. K., Gaur, A., Saddikuti, V., & Rastogi, A. (2017). Structural equation model (SEM) – neural network (NN) model for predicting quality determinants of e-learning management systems. *Behaviour & Information Technology*, *36*(10), 1053–1066. https://doi.org/10.1080/0144929x.2017.1340973.

Shet, S. V., & Pereira, V. (2021). Proposed managerial competencies for Industry 4.0 – Implications for social sustainability. *Technological Forecasting and Social Change*, *173*, 121080. https://doi.org/10.1016/j.techfore.2021.121080.

Sim, J.-J., Tan, G. W.-H., Wong, J. C. J., Ooi, K.-B., & Hew, T.-S. (2014). Understanding and predicting the motivators of mobile music acceptance – A multi-stage MRA-artificial neural network approach. *Telematics and Informatics*, *31*(4), 569–584. https://doi.org/10.1016/j.tele.2013.11.005.

Smith, W. L., Schallenkamp, K., & Eichholz, D. E. (2007). Entrepreneurial skills assessment: an exploratory study. *International Journal of Management and Enterprise Development*, *4*(2), 179–201.

Sivathanu, B., & Pillai, R. (2018). Smart HR 4.0 – how industry 4.0 is disrupting HR. *Human Resource Management International Digest*, *26*(4), 7–11. https://doi.org/10.1108/hrmid-04-2018-0059.

SIGKDD. *International Conference on Knowledge Discovery and Data Mining*, pp. (1435–1444), doi: 10.1145/2783258.2783266.

Society for Human Resource Management. (2019). *The Global Skills Shortage – Bridging the Talent Gap with Education, Training and Sourcing.* SHRM.

Souza, F. F. D., Corsi, A., Pagani, R. N., Balbinotti, G., & Kovaleski, J. L. (2021). Total quality management 4.0: Adapting quality management to Industry 4.0. *The TQM Journal*, (ahead-of-print). https://doi.org/10.1108/tqm-10-2020-0238.

Tan, G. W.-H., Ooi, K.-B., Chong, S.-C., & Hew, T.-S. (2014). NFC mobile credit card: The next frontier of mobile payment? *Telematics and Informatics*, *31*(2), 292–307. https://doi.org/10.1016/j.tele.2013.06.002.

Tehseen, S., Ahmed, F. U., Qureshi, Z. H., Uddin, M. J., & Ramayah, T. (2019). Entrepreneurial competencies and SMEs' growth: The mediating role of network competence. *Asia-Pacific Journal of Business Administration*, *11*(1), 2–29. https://doi.org/10.1108/apjba-05-2018-0084.

Udemy. (2020). *2019/2020 Skills Gap Report.*

Volery, T., Mueller, S., & von Siemens, B. (2013). Entrepreneur ambidexterity: A study of entrepreneur behaviours and competencies in growth-oriented small and medium-sized enterprises. *International Small Business Journal: Researching Entrepreneurship*, *33*(2), 109–129. https://doi.org/10.1177/0266242613484777.

Wang, Y.-F. (2013). Constructing career competency model of hospitality industry employees for career success. *International Journal of Contemporary Hospitality Management, 25*(7), 994–1016. https://doi.org/10.1108/ijchm-07-2012-0106.

Wong, L. W., Leong, L. Y., Hew, J. J., Tan, G. W. H., & Ooi, K. B. (2020). Time to seize the digital evolution: Adoption of blockchain in operations and supply chain management among Malaysian SMEs. *International Journal of Information Management, 52*, 101997. https://doi.org/10.1016/j.ijinfomgt.2019.08.005.

World Economic Forum. (2018). *The Future of Jobs Report 2018*. WEF. http://www3.weforum.org/docs/WEF_Future_of_Jobs_2018.pdf.

World Economic Forum. (2020). *The Future of Jobs Report 2020*. WEF. http://www3.weforum.org/docs/WEF_Future_of_Jobs_2020.pdf.

Zhang, J., Yu, P. S. and Lv, Y. (2015), Organizational chart inference, *Proceedings of the 21th ACM*.

APPENDIX

Sr. No.	Author	Purpose	InOrdination
1	Hecklau et al. (2017)	Examining the impact and consequences of I4.0 technologies on employee competencies.	198.00
2	Prifti et al. (2017)	Analyse the employees' competencies in I4.0 in higher education institutions.	172.00**
3	Benešová and Tupa (2017)	To study the students' competencies and education in I4.0.	150.00
4	Grzybowska and Łupicka (2017)	The purpose of this study is to examine the competencies need to cope with I4.0 challenges.	140.00**
5	Popkova and Zmiyak (2019)	The experimental study identifies the priorities of competencies (social & technical) during training in I4.0.	133.00
6	Kortuem et al. (2013)	The experimental study examines the role of Internet-of-Thing's technologies in student education.	117.00
7	Schallock et al. (2018)	The objective of this research is to address the growing demand for future skills of production staff.	113.00**
8	Enke et al. (2018)	To study and examine the competencies required for I4.0 that enable the integration of lean management and I4.0 technologies.	105.00**
9	Eguchi (2014)	The purpose of this case study research is to understand the role of robotics education to promote digital skills.	103.00*
10	Demartini and Benussi (2017)	To understand the role of web 4.0 and I4.0 in education industry to promote the digital competencies.	101.00

(*Continued*)

Sr. No.	Author	Purpose	InOrdination
11	Sivathanu and Pillai (2018)	To present the conceptual framework for human resources management in I4.0.	99.00*
12	Meyer et al. (2015)	To analyse the highly relevant manufacturing-related competencies required for today and future employees.	97.00*
13	Babatunde (2020)	To study the soft and hard quality management competencies necessary for I4.0.	88.00
14	Gocmen & Coskun (2019)	To identify the essential elements of engineering education needed to adapt I4.0 vision and technologies.	87.00
15	Kipper et al. (2021)	To study, identify and mapping the essential employees' competencies for I4.0.	85.00
16	Hecklau et al. (2017)	To systematically examine the employee competencies needed within I4.0.	85.00**
17	Kazancoglu and Ozkan-Ozen (2018)	To examine the employees' competencies in I4.0 and develop a structural competency model.	84.01
18	Caruso (2018)	To understand the association between technologies, innovation, and social changes in manufacturing firms.	83.00
19	Shet and Pereira (2021)	To study and proposing the essential managerial competencies required for I4.0 to promote social sustainability.	80.01
20	Flores et al. (2020)	To address and propose human competencies needed in digitalization.	80.01
21	Saniuk et al. (2021)	To study the key areas of knowledge and skills necessary to implement I4.0 concept.	80.00

Note: *Scopus database, **Google scholar.

2 Industry 4.0 and Emerging Trends in Techno-Entrepreneurship

Bhakti Parashar
Vellore Institute of Technology-Bhopal University

Ravindra Sharma
Swami Rama Himalayan University

Zaiba Ali
Princess Nourah Bint Abdulrahman University

Shivani Agarwal
KIET Group of Institutions

CONTENTS

DOI: 10.1201/9781003217084-2

2.1 INTRODUCTION

An entrepreneur is a person who works for himself for a period of time before launching a new business (Markussen and Røed, 2017), since innovation is becoming a more important area for business in a more competitive world with a growing need for perfection. It is widely acknowledged as an essential component in achieving customers' or target audiences' actual expectations, as well as a source of competitive advantage (CA), performance, and development (Day, 1994, Peteraf, 1993). However, in this era of Industry 4.0, it is nearly impossible for innovation to thrive without technological advancement. At a macro level, innovation prospects and capabilities point to a technological revolution that will allow for the establishment of a new techno-economic paradigm, new linked technologies and commodities (Perez, 2009; Singh et al., 2019; Sharma et al., 2021a,b).

The term "Industry 4.0" refers to the ongoing change in the global industrial industry. Large companies, in particular, have swiftly acknowledged Industry 4.0's challenges and are now concentrating their efforts on deploying the essential enabling technologies. Firms, individuals, and virtual communities may now distribute and coordinate user knowledge at unprecedented levels because of the internet's advent and broad usage (Sharma et al., 2021a).

Smart enterprises include automated systems and digital enablers that allow for machines to connect with one another and with the factory's overall systems via the internet of things, and these skills are in high demand across businesses as they strive to keep their manufacturing units competitive in an increasingly technological world (Renjen, 2021). That is why concept of technology entrepreneurship is a noteworthy and current area of management science research. While entrepreneurship theory and innovation theory both have well-established research results, connecting them to the larger phenomenon of technology entrepreneurship remains a difficulty. Technology entrepreneurship has the ability to bring management and entrepreneurship concepts closer together.

People's capacity to effortlessly share their knowledge resources has paved the way for new business prospects. Many traditional entrepreneurship and innovation projects are developed from an internal, supply-side perspective, neglecting the potential contributions of nonfirm actors, such as users, and the knowledge resources they might bring to the innovation and operations processes (Sharma and Rana, 2021). One of the most pressing concerns confronting businesses today is selecting the appropriate strategy for gaining CAs in the age of Industry 4.0. This is required for their long-term commercial viability.

In all types of enterprises, CA is a significant notion in strategic management and a key component in the process of enhancing competitiveness. Competitiveness is a strategic aspect in terms of long-term functioning and development, as are the efficiency, dynamism, and flexibility of the investigated firm, according to their type (Adamik and Nowicki, 2018).

The intelligent manufacturing system (IMS) is a next-generation manufacturing system that transforms a traditional production system into a smart system by applying new models, forms, and procedures. An IMS provides a fully integrated human–machine production system in the Industry 4.0 era by delivering collaborative, customizable, adaptable, and reconfigurable services to end-users through the internet (Prakash et al., 2021; Ammar et al., 2021).

Despite being most technologically advanced, implementation and validation of Industry 4.0 are jeopardized without a thorough understanding of and judicious application of productive methods (Rana and Sharma, 2021a,b). Finding the proper method to gain CAs in the era of Industry 4.0 is one of the most challenging tasks facing today's company leaders, and it is essential for their long-term economic sustainability (Adamik and Nowicki, 2018; Sharma et al., 2019). As a result, enterprises must begin with identifying and gradually resolving requests immediately in order to improve the efficiency of their market operations and grow their CA. Understanding the unique characteristics and requirements of the Industry 4.0 era should be the first step in the process of technological advancement in entrepreneurship. Digitalization and big data analytics—sometimes known as datafication—are pervasive in all aspects of life, resulting in new methods of working, communicating, and cooperating (Loebbecke and Picot, 2015). The evolution of a company's strategy is substantially influenced by digitalization at present (Holotiuk and Beimborn, 2017).

2.2 HISTORY AND DEVELOPMENT OF ENTREPRENEURSHIP

The word "entrepreneur" derives from the French word "entrependre," which means "to undertake." According to Carlen (2016), the term "entrepreneurship" did not become popular until the late 1800s, when Jean-Baptist Say created it. The title undertaker has long been used in English-speaking countries, but it was not until the mid-nineteenth century that the term entrepreneurship, which has French roots, became the official designation (Joško et al., 2019). Entrepreneurship has piqued the curiosity of management professionals and social scientists since the 1980s.

The subject is becoming increasingly popular, particularly among business schools (Cooper, 2003). The notion was crucial in the development of business history as a distinct academic discipline. Economic historians have been chronicling how economies have evolved through time since the middle of the 19th century, criticizing the static assumptions of classical and neoclassical economic philosophy. The early historicism was concerned with the evolution of capitalism and industrialism's institutions (Hodgson, 2001). Several terminologies were used to try to describe the phrase, one of the most prevalent being "adventurer," which may be translated as "one who discloses something."

The term "adventure" also has been used to describe a business activity in which an entrepreneur travels for the purpose of financially benefiting himself and his family over the course of his many years of labor. The most significant contribution to the development of a scientific and professional approach to entrepreneurship has been made by

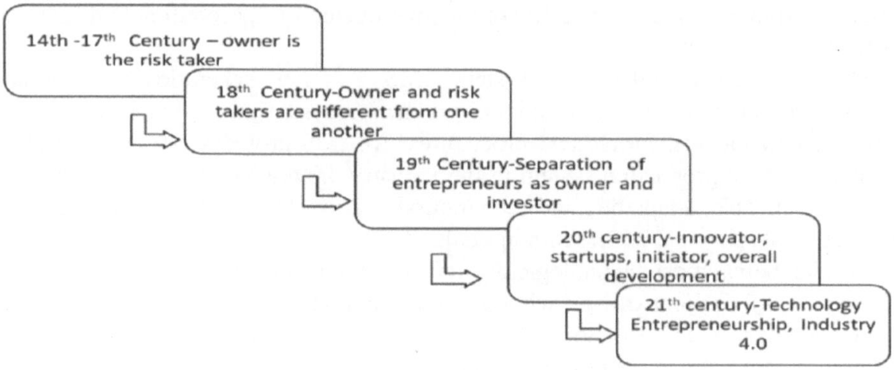

FIGURE 2.1 Development of entrepreneurship. (*Source:* Authors own.)

Peter Drucker. Drucker (2015) agreed with Say's definition of entrepreneurship from the 1800s as "an activity that shifts resources from low productivity and earnings to higher productivity and profits" but suggested that the term has been defined more precisely.

After watching a range of entrepreneurial actions, Drucker came to the conclusion that not every act of producing new value or transferring money to the 49th state is an act of creating new value. On a worldwide scale, entrepreneurial growth phases may be classified into five main periods. Since the birth of entrepreneurship and the expansion of capitalism is essentially a European phenomenon, we must frame and emphasize that global practice is strongly tied to European practice (Figure 2.1).

2.2.1 Earliest Period

The early entrepreneurs were traders and merchants. New Guineans traded obsidian, a black volcanic glass used to make hunting arrowheads, for other essentials about 17,000 BCE, marking the first known instance of people bartering. These forerunners in the corporate world swapped one set of goods for another. The earliest animals were domesticated approximately 15,000 BCE, while the first plants were domesticated around 10,000 BCE. This achievement in agriculture was crucial to the human race's progress.

At that time, agriculture enabled humans to build larger stationary groups and cities (civilization's foundations), allowing for human knowledge to spread and progress. It has transformed civilization, allowing for permanent settlements to replace nomadic ones and laying the framework for human population to increase from 15 million to 7 billion in the millennia ahead.

As the last Ice Age ended in approximately 8,000 BCE, the poles melted, causing sea levels to rise and separating Siberia and North America. This split separated two human civilizations for about 10,000 years, until European explorers arrived in the Americas in the 15th century.

2.2.2 14th–17th Centuries

This is the first age of entrepreneurship, when a population explosion occurred in about 1470, resulting in a spike in cities, marketplaces, and trade volume. The guild system grew in popularity, and the concept of a business as a faceless entity with its own identity independent from its owner gained acceptance. Production and trade did not require significant long-term investment in the past. A family and friend collaboration might be able to gather enough funds (Hunt and Murray, 1999). Long-distance trade, on the other hand, necessitated a growth in size and risk-bearing ability in the 16th and 17th centuries, necessitating investment from a diverse group of investors beyond their usual circles. This resulted in a two-pronged problem: how to foster long-term financial commitment in the company while also avoiding the risk of expropriation by private and public actors.

Banking, which originated with the Ancient Mesopotamians, has grown to unprecedented heights of complexity. As silver imports from the new world spurred an increased trade, bookkeepers developed standardized ways for keeping track of a firm's finances based on Luca Pacioli's accounting triumphs. By raising money, taking risks, and supporting economic growth, early entrepreneurs known as merchants and explorers began to generate capital, take risks, and drive economic progress. The proprietor was a risk-taker, and this period may be seen as the beginning of capitalism. A war of ideas raged to explain where money originated from and how the market operated with the introduction of a sophisticated marketplace and a capitalist economy.

Mercantilism was a popular idea between 1400 and 1700, where a country's prosperity was largely determined by how much treasure and gold it could get, as well as how much more it sold than it imported. Trade and competition were discouraged, while monopolies and tariffs were supported. They were, however, entirely incorrect.

2.2.3 18th Century

Nonetheless, new schools of thought evolved in the 18th century that emphasized business as a source of prosperity rather than the mercantile mentality of gold hoarding. Traditional business historians, like general economic historians, sought to relate the rise of capitalism to improvements in competitive marketplaces and the establishment of "open societies." From this vantage point, war was regarded as a constant source of market disruption caused by government interference and as the preferred arena of rent seeking over entrepreneurial behavior (Baker, 1971).

This opinion was shared by Adam Smith, the first to recognize and explain the heart of the marketplace. In his famous 1776 essay "An Inquiry into the Nature and Causes of the Wealth of Nations," he slayed the mercantilist monster. Self-interest, according to Smith, operates as a directing force toward the labor that society demands. The owners were separated from the risk-takers.

What ultimately defined the 18th-century entrepreneur was the interrelationship of various fields, rather than specialization in one form of work. In a world rife with uncertainty, as NualaZahedieh demonstrated, the best approach to hedge these risks was to intertwine multiple business lines (Parrott, 2012). Thomas Newcomer's steam engine was created in Devon, England, in 1712, kicking off the Industrial Age. Things did not really take off until 1763, when James Watt invented the steam engine, which replaced muscle action with pistons.

2.2.4 19th Century

The Industrial Revolution was a turning point in human history of world business. The late-18th-century Industrial Revolution transformed 19th-century commerce and industry, resulting in substantial advancements in transportation, industrial production, communications, and energy. From both an economic and cultural aspect, the outcome was a significant shift in society. For entrepreneurs with vision and courage, this rich atmosphere produced amazing business opportunities.

The marine business was at its peak in the middle of the 19th century, when petroleum replaced whale oil, indicating the limited nature of opportunity (Hunter, 2013). The French economist Jean-Baptiste Say came up with a basic definition of entrepreneurship in the starting of the 19th century, claiming that entrepreneurship "improves the productivity and efficiency of economic resources" (Ghicajanu, 2015).

With the help of a government that fostered risk-taking and minimal regulatory intervention, 19th-century entrepreneurs established an American system of capitalism that is now admired across the world. When finding dynamic business innovators from the 19th century, it is intriguing to see what traits they have in common with successful entrepreneurs from earlier eras.

2.2.5 20th Century

Protectionism and national self-interest made the initial part of the 20th century a wreckage for progress and development. In the course of less than 30 years, two world wars occurred. Britain remained a colonial power, Japan was rapidly establishing itself as the East's power, South-East Asia remained underdeveloped, and Roosevelt's quasi-socialist New Deal policies hindered American commerce (Schweikart and Doti, 2010). Despite the fact that the Second World War aided the development of new technologies, it was only after the war that entrepreneurs reappeared in what may have been the world's quickest 30 years of expansion. A slew of new industries based on newly developed technologies sprang up, dramatically altering society (Hunter, 2013).

Many 19th-century entrepreneurs, such as industrial laboratories and skunk works, set the tone for how people and corporations would develop and innovate later in the 20th century. Automobiles, airplanes, steel, petroleum, electric power, communications, media entertainment, and processed goods were among the businesses that expanded into large enterprises as the century progressed. In the spirit of Schumpeter's idea of creative destruction, entrepreneurship and innovation moved society forward, and this was best done when it was fostered (Hunter, 2013).

2.2.6 21st Century

In the new global economy of the 21st century, entrepreneurship's position and functions have risen in importance and complexity, creating new issues. The 4.0 Industrial Revolution is centered on globalization, innovation, trade liberalization, and information and communication technology, offering the trademarks of the 21st-century economy and encouraging a proactive approach to the worldwide expansion of the entrepreneurial ecosystem (Mason and Brown, 2014).

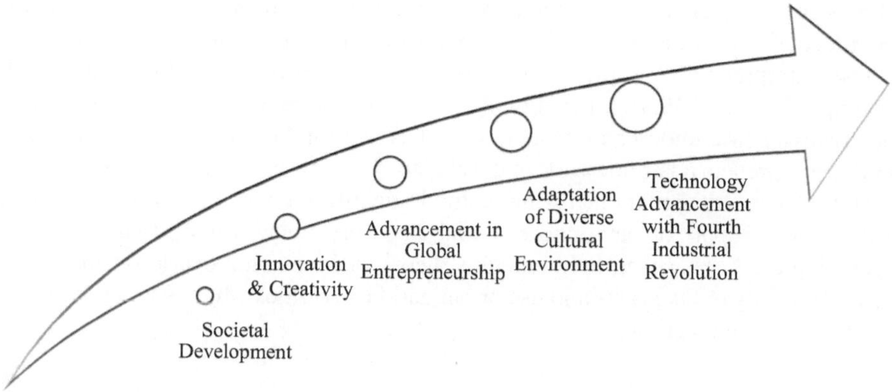

FIGURE 2.2 Developmental changes in entrepreneurship in the 21st century. (*Source:* Authors own.)

2.3 DEVELOPMENTAL CHANGES IN ENTREPRENEURSHIP

Organizations in the knowledge society work in a global economy that is characterized by fierce rivalry as well as economic interdependence and collaboration. Thousands of jobs have been lost as a result of globalization of goods and service production, particularly in manufacturing, owing to automation or transfer to richer nations (Anderson, 2008; Levy and Murnane, 2012), and this was only possible with the development of entrepreneurship development.

Some of the developmental changes in entrepreneurship in the 21st century are represented in Figure 2.2.

2.3.1 Societal Development

Individual involvement in community life is accompanied by entrepreneurship. The development of entrepreneurial abilities improves social standing by allowing individuals to obtain financial independence and social respect. Furthermore, the consequences are felt at the local and national levels, as a result of economic growth and, implicitly, an increase in the standard of living.

The entrepreneurial ecosystem is always dynamic in nature on the contrary science and technology have progressed beyond the stage of transformation and have become a crucial component of global economic and social growth. In the modern economy, entrepreneurship is dependent on a consistent and purposeful degree of innovation. From the global to the local level, the integration of innovation has become a constant goal for economic efficiency. Entrepreneurship alters the trajectory of a country's national economy. The level of entrepreneurship in a society determines its overall well-being.

2.3.2 Innovation and Creativity

In an increasingly competitive world where there is a growing push for perfection, innovation is becoming a more critical industry for organizations. It is widely acknowledged as an essential component in achieving customers' or target audiences' actual expectations, as well as a source of CA, performance, and growth.

The entrepreneur's profile is described by three primary features in Joseph Schumpeter's (1883–1950) innovation thesis on entrepreneurship: vision, innovation, and creativity (Mehmood and Alzoubi, 2019). According to this theory, an entrepreneur's entrepreneurial framework is developed when he or she makes a new product, introduces a new manufacturing method, identifies a new market for a product, locates a new supply of raw materials, and discovers a new way of doing things or organizing them. However, Schumpeter's theory of innovation overlooks the entrepreneur's ability to take risks and use organizational abilities while placing an excessive emphasis on creativity.

2.3.3 Advancement in Global Entrepreneurship

Globalization has eliminated the national borders and rendered geographic location irrelevant, creating an environment in which gaining a global perspective is one of the primary goals of entrepreneurship, with a global perspective providing the benefit of facilitating the integration of global economic opportunities.

The emerging world economy's constant economic expansion has increased nation-to-nation interdependence and improved the links between production and marketing. The nascence of the global entrepreneur, with the ability to adopt a global perspective in the pursuit of entrepreneurial ventures, is at the heart of this transition. In order to include multidimensional relationships and complicated social, cultural, economic, and political realities into a global plan, the modern entrepreneur must broaden his mentality.

2.3.4 Adaptation of Diverse Culture Environment

A complete approach to workforce inclusion, in which all employees are treated fairly and given equal opportunities and rewards, is referred to as a global perception (Rana and Sharma, 2017). It is all about cultivating human synergy, where the final product transcends individual contributions owing to careful coordination among a varied and pluralistic workforce.

A holistic approach to workforce inclusion entails treating all employees equally and providing them with equal opportunities and benefits. It is all about encouraging human synergy, where the final product transcends individual contributions owing to conscious coordination among a varied and pluralistic workforce (Rana and Sharma, 2021a,b).

2.3.5 Technology Advancement with the Fourth Industrial Revolution

Entrepreneurial opportunity study looks at how businesses employ technological advancements to create new processes, goods, services, markets, or organizational structures. At a macro level, the innovation opportunity space refers to a technological revolution that allows for the formation of a new techno-economic paradigm that in turn allows for the development of new types of connected technologies and products, while at a micro level, it refers to the market potential perceived by entrepreneurial R&D managers in design organizations (Perez, 2009; Meyer and Tucker, 2010).

2.4 ENTREPRENEURSHIP AND INDUSTRY 4.0

From product conception and purchase through production, delivery, and service, industrial revolutions have constantly contributed to entrepreneurial development through technology. Across the board, Industry 4.0 vertically integrates and digitizes corporate operations. All data on operation processes, process efficiency, quality control, and operation planning is real-time and networked.

2.4.1 Industrial Revolutions and Entrepreneurship Development

The new economy has come a long way since its inception at the dawn of the Industrial Revolution in the 18th century. For decades, animal labor has been utilized to make or use most of what we have, including firearms, tools, food, clothes, and homes. This condition improved with the development of industrial methods in the late 18th century. After that, the growth of Industry 1.0 was a quick ascension to the next manufacturing period, the fourth era. Figure 2.3 shows the history and development of industrial revolutions. As seen in Figure 2.3, the Industrial Revolution did not begin today; rather, it began in the 18th century, which is also known as Industry 1.0, followed by Industry 2.0, 3.0, and finally Industry 4.0, which is given in detail below.

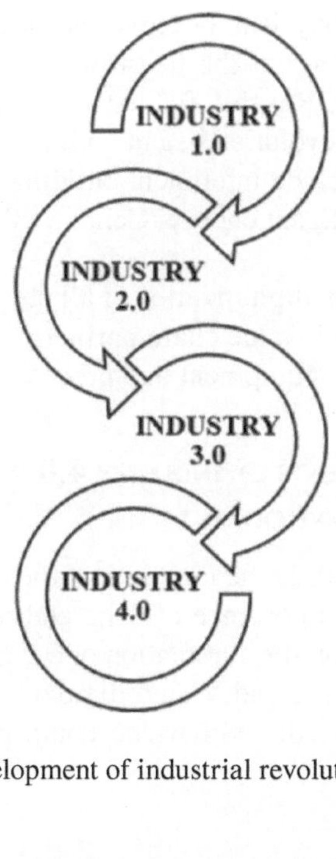

FIGURE 2.3 History and development of industrial revolutions.

- **Industry 1.0:** The advent of water power and energy output in the 18th century signified the commencement of the First Industrial Era. In the same amount of time, the automated version, which previously made threads on single spinning wheels, produced eight times the length. We were previously familiar with its steam engine. What was employed for manufacturing purposes saw the greatest increase in human productivity. Instead of spinning muscle-driven looms, steam engines might be employed (Antony et al., 2005).
- **Industry 2.0:** The Second Industrial Revolution, sometimes known as Manufacturing 2.0, began at the turn of the 20th century. The discovery of electrical energy technology launched this revolution. Electrical technology has also been employed as a key source of power (Duraković, 2011; Goh, 2011). Pigs were hung from conveyor belts at a Chicago slaughterhouse, and each butcher barely performed half of the job of killing a cow. Henry Ford (1863–1947) applied the concept of mass processing. These principles were implemented in the manufacturing of automobiles by Henry Ford, and the result was a tremendous improvement.
- **Industry 3.0:** In the 1970s, the Third Industrial Revolution began with the semiautomation of operations utilizing memory-programmable controllers and computers (Thomas and Lewis, 2007). Industry 3.0 connects Henry Ford's desire for efficiency with the sophisticated processes that are emerging as part of Industry 4.0. At Ford, not only were operations optimized but also automation made critical aspects of the manufacturing process safer and more efficient.
- **Industry 4.0:** Industry 4.0 was presented by Germany in 2011 as a way to develop a new model of German economic strategy centered on high-tech businesses. Cyber-physical systems, the internet of things (IoT), and the Internet of Services are at the heart of the fourth technological revolution (Lasi et al., 2014; Rakshit & Sharma, 2020; Ning and Liu, 2015). This Fourth Industrial Revolution began with industry (smart factories) and has now expanded to include intelligent buildings, intelligent mobility and transportation, and intelligent energy (Cellary, 2019).

Industry 4.0 focuses on the total digitalization of all physical assets and their integration into digital ecosystems with value chain partners, whereas Industry 3.0 focused on the automation of particular equipment and processes.

2.4.2 Implication of Elements of Industry 4.0 in Entrepreneurship Development

Industry 4.0 (Górka et al., 2020) is the outcome of the digitalization of manufacturing processes, the increasing importance of data, and the use of cyber and physical systems. Industry 3.0 focused on the automation of certain equipment and processes, whereas Industry 4.0 stresses the end-to-end digitization of all physical assets and integration into digital ecosystems with value chain partners. Industry 4.0 offers benefits based on the seamless collection, analysis, and transfer of data, as well as the integration of a variety of new technologies to create value.

Many industrialists, entrepreneurs, organizations, research centers, and universities are currently focusing on the fourth Industrial Revolution, but most academic experts believe the term Industry 4.0 is vague, making it difficult for manufacturing companies to grasp the phenomenon quickly and determine the necessary steps for the transition to Industry 4.0. In 2011, the German government coined the phrase "Industry 4.0" as a concept and strategic goal for Germany and other industrialized countries (Kumar et al., 2020).

In the following, we present the explanation of these essential elements of Industry 4.0 and discussion on how these elements contribute to entrepreneurship development. Figure 2.4 is the diagrammatic representation of the essential elements of Industry 4.0.

1. **Big Data:** Interconnected heterogeneous things can create a large volume of data of many types (Bortolini et al., 2017). The term "big data" describes a large volume of structured, semistructured, and unstructured information (BD). Keeping and analyzing these data would take far too much time and money to provide the same benefit (Qi and Tao, 2018). In the Internet of Everything age, connecting more physical items to the internet and employing a generation of new technologies are two approaches to give value potential to organizations. BD is defined by data gathering and storage, but data analysis is its basic characteristic, and without it, BD is meaningless.

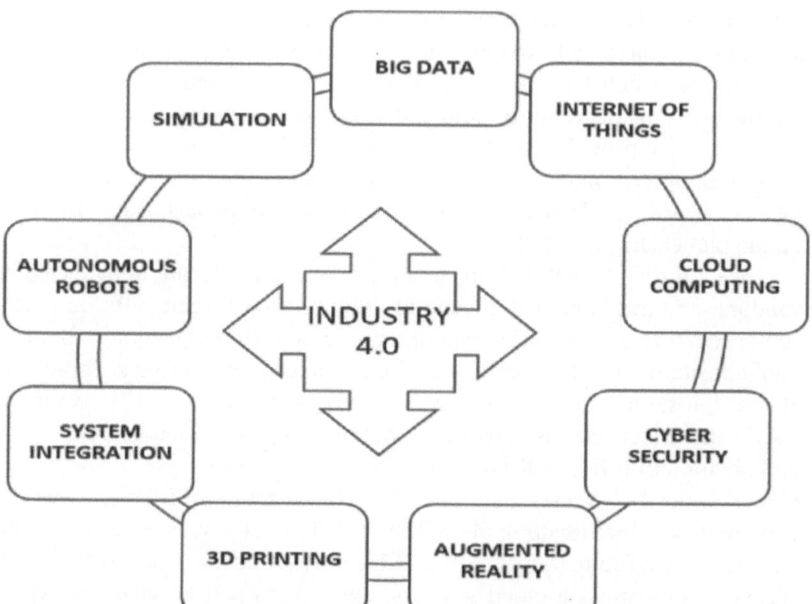

FIGURE 2.4 Elements of Industry 4.0. (*Source:* Authors own.)

BD can help managers with decision-making and/or problem-solving related to operations throughout the product life-cycle (Tao et al., 2018), resulting in cost-effective and fault-free process operation (Yin and Kaynak, 2015), and assisting managers with decision-making and/or problem-solving related to operations (Tao et al., 2018). The use of BD gives us a CA since it allows us to create value-added goods (Cheng et al., 2018).

2. **Internet of Things:** The IoT is the combination of two concepts in the IT world: "internet" and "things." The "Internet" is a network of networks. TCP/IP is a global technology that connects interconnected computer networks to serve people all over the world. The IoT is currently widely employed in a variety of fields, such as transportation, healthcare, and utilities (Sezer et al., 2018). A network of Things-to-Things, Things-to-Humans, and Humans-to-Humans links to the internet inside the IoT. Individually identifiable entities exchange data within this network (Sadiku et al. 2017; Sharma et al., 2021d; Choi and Chung, 2017; Sadiku et al., 2017).

   Today's entrepreneurs and industries are facing new market requirements and competitiveness; thus, they need to know what is known as Industry 4.0, which is a profound revolution. Integration of artificial intelligence (AI) with new disruptive technologies such as the Industrial Internet of Things (IIoT) and their application is also performed. As a result, the IoT enables people and objects to connect at anytime, anywhere, with anything and anybody, preferably via any path/network and using any service (Sezer et al., 2018).
3. **Cloud Computing:** Cloud Computing may be used by companies who want to outsource their IT (CC). CC for SMEs is defined as a resource pooling solution that offers quick elasticity and metered service, as well as self-service on demand and expanded network access (Assante et al., 2016). The cost-cutting benefits of using CC include the direct and indirect costs of removing IT infrastructure from the organization, the resource rationalization service provided by dynamically scalable users who consume only the computing resources they actually use, and portability when using any internet-connected device, such as mobile phones or tablets, from anywhere on the planet (Branco et al., 2017).

   The cloud can be public (typically on a data center location, managed by vendors, and available to the general public), private (generally on a data center location, managed by vendors, and closed to the general public), or public (usually on a data center location, managed by offering specific benefits at the same site as the organization, hybrid: community, or a mix of public and private clouds: a hybrid of public and private clouds) (Alqaryouti and Siyam, 2018; Xu, 2012).
4. **Cyber Security:** Every year, more devices connect to the internet, a worldwide network. Inanimate items will be the dominant source of data in the not-too-distant future (Sergey et al., 2019). IoT, virtual environments, remote access, data stored on cloud systems, and other open alternatives expose people and enterprises to greater risks. Because the enterprise boundaries are ambiguous and dissolving, the danger scenario becomes a reality (Sharma et al., 2021e; He et al., 2016).

Safety communications at each stage of the manufacturing process, as well as safety interoperability among facilities, are key components of supply chain value. I4.0 technologies must allow for the creation of a secure cyber environment with cyber security measures in place. A cyber-attack can cause manufacturing activities to be interrupted, resulting in financial losses for businesses. On the other hand, cyber-attacks on systems that need safety operations are a huge concern, posing a substantial risk to the operators' safety (Tsuchiya et al., 2018).

5. **Augmented Reality:** Augmented reality is used in a variety of industries, including entertainment, marketing, tourism, surgery, logistics, manufacturing, and maintenance (Palmarini et al., 2018), its usage has recently moved to various production industries as a rapidly emerging technology (Mourtzis et al., 2017). The application of augmented reality (AR) in industrial processes for simulation, assistance, and guiding has been shown to be an effective tool for solving problems (Rentzos et al., 2013). When compared to traditional hardware, AR has a significant benefit in terms of buying cost, which can be modest or even zero. In other circumstances, the see-through glasses component might be costlier (Horejší, 2015). Another key advantage of assets in the industrial setting, according to Blanco and others (Blanco-Novoa et al., 2017), is that because AR delivers dynamic real-time information, it may reduce the majority of paperwork.
6. **3D Printing:** 3D printing technology may be used for a number of purposes and at different phases of the entrepreneurial process. To evaluate the advantages of 3D printing for company owners, we take into account the technology's specific use and the stage of the production process where it is utilized. The potential for 3D printing to change entrepreneurship and the manufacturing process is apparent. A growing number of major players in other industries, such as aerospace and automotive, have adopted 3D printing technology to directly manufacture end-use parts and products—Airbus, Ford, and General Electric are just a few of many companies that use 3D printing extensively—and this is a very good sign for entrepreneurs to adopt this technology and grow their business.
7. **System Integration:** Technological innovation in production systems needs organizations with flexible operational processes and efficient information management to establish organizational synergy and deliver CAs inside the production system and value chain. Divergences in management have been reported, affecting operations primarily of businesses with an Industry 4.0 mindset; this has created a gap for those who do not meet these criteria, demonstrating the rapid transformation and accomplishments achieved by businesses that have adapted their processes to the new system. The systems are looked at as a whole, with the productive flow in mind. The arrangement and management of physical items, as well as the construction of linkages with information systems, are proposed to undergo structural modifications (Dombrowski and Wagner, 2014; Singh et al., 2019; Fielt, 2013). The development and execution of a company's operations are referred to as vertical flow, and it comprises critical variables, including organizational structure,

human factor, departmental linkages, technological level, and managerial level. External relations, supplier and customer network integration, information and management systems, and more are all part of the horizontal flow (Lee et al., 2014; Vyas and Nijholt, 2016).

8. **Autonomous Robots:** The manufacturing paradigm of entrepreneurs is transitioning from mass to tailored production, necessitating the usage of reconfigurable automation technologies such as robots. This proclivity has an influence on manufacturing businesses' production systems since it leads to production adaption for a larger range of goods, with a preference for one-batch sizes. In order to attain the requisite level of flexibility, robots are increasingly essential in industrial operations (Pedersen et al., 2016). When microprocessors and AI are used to make goods, services, and machines smarter, new capabilities in computing, communication, control, autonomy, and sociality emerge. Robots with AI that are adaptable and versatile can help in the manufacture of a wide range of products, cutting production costs (Salkin et al., 2018).
9. **Simulation:** The use of computer simulation to better understand the dynamics of corporate systems is becoming more common (Rodic, 2017) and it is a necessary and powerful instrument for the effective deployment of digital manufacturing (Mourtzis et al., 2014), Current manufacturing sector concerns can be addressed by this technology (Zúñiga et al., 2017), which deals with system complexity and components of uncertain issues that standard mathematical models can't answer (Lachenmaier et al., 2017). The benefit of simulation in a customized product manufacturing environment is astounding and clear. Cost savings, shorter development cycles, and higher product quality are all benefits of simulation modeling (Rodic, 2017). Industrialists and entrepreneurs have been employing modeling and simulation to examine their operations and support decision-making (Shao et al., 2014). Simulation technologies have previously demonstrated their efficacy in the solution of a number of practical real-world challenges in the manufacturing industry (Negahban and Smith, 2014).

2.4.3 Significant Role of Industry 4.0 in Various Fields of Entrepreneurship

The Fourth Industrial Revolution encourages companies to align their business models, which consider information as a CA and are fueled by data collecting, utilization, and sharing opportunities. Organizations' fundamental objectives, the wellspring of advantage, goods, and administration portfolio are all outlined in plans of action. A transformation strategy is critical for maintaining a focused edge in the concept of Industry 4.0.

The cultivation of respect for the organization and its partners is at the center of new action plans.

Furthermore, modern innovation arrangements change workers' roles from administrators to problem-solvers, allowing for cooperation throughout the whole value chain and increasing client engagement in product, service, and design development (Jung et al., 2017; Arnold et al., 2016).

Although there are several benefits of Industry 4.0 to entrepreneurship, the following are the most important ones.

- **From Automation to Autonomous:** Manufactured products, like autonomous machines, could be autonomous items with self-reliant or self-governing capabilities based on connectivity and computational capacity. Machines that can learn to conduct themselves can reduce expenses while also producing items that are more obedient to customer standards (Sipsas et al., 2016). Each equipment and its process in a cognitive factory is outfitted with cognitive skills similar to human behaviors, allowing for industrial settings to adapt to changes flexibly and independently (Petrelli, 2017).
- **Virtualization:** Makers who follow Industry 4.0 will create a "virtual twin" of a brilliant industrial facility to enhance current processes and goods, cut item development and demonstration time, construct a manufacturing procedure, and therefore, reduce the chance of benefiting from new things. (Brettel et al., 2014)
- **Smart Life-Cycle Data:** This information should ideally be shared across all product life-cycle stakeholders to better product life-cycle management (Kurilova-Palisaitiene et al., 2015). However, the flow of product information remains mostly unestablished due to insufficient data extraction, information loss during product transfer between stakeholders, underdeveloped platforms to facilitate information sharing, and other legislative constraints. As a result of data circulation inefficiencies, product life-cycle management and service quality have degraded. In order to bring these things to an "as new" state, remanufacturers must replicate product knowledge that existed at the time of product development.
- **Logistics:** In the logistics industry, the barcode system is employed, and RFID is one of the most trustworthy sources they use these days to assure smooth operations (Lee et al., 2015). One of the most significant advancements in the business is the ability to track freight from the time it leaves the warehouse until it reaches the intended receiver (Gilchrist, 2016). The precision and speed of RFID technology are its most crucial features. In a large company, this method may be utilized for both internal audits and external training.
- **Smart Future Factories:** The complexity issue connected with remanufacturing processes is readily handled by smart factories, which allow for significant flexibility and small batch size production. Industry 4.0 is often referred to as "smart factories". To create "smartness," electronic hardware/software and the networking of industrial resources are employed. Auxiliary hardware and software, such as RFID tags, barcodes, laser markers, sensors, and communication infrastructure, will be included in the factory, allowing for machines to communicate utilizing intelligent analytics.
- **A Competitive Economy with a High Wage Rate:** Due to the twin strategy of Industry 4.0, Germany has been able to establish itself as the key provider as well as the primary market for Industry 4.0 setups. Industry 4.0 cannot, in any case, be viewed as an orderly or IT-related test for major businesses.

Dynamic innovation is also expected to have broad governmental implications, as it allows for the growth of new organizations and corporate models, hence driving greater levels of representational commitment.

2.5 COMPLEXITIES WITH INDUSTRY 4.0 IN ENTREPRENEURSHIP

Globalization has created a networked world that has an impact on both corporate and personal life. They have also brought about a spate of developments in the industrial sector (Bauer et al., 2015). We are on our way to creating a ubiquitous knowledge society replete with smart and autonomous technology.

Two major obstacles to Industry 4.0 deployment are a lack of trained labor and the need to retrain individuals to match changing conditions. (Adolph et al., 2014; Erol et al., 2016; Shamim et al., 2016; Karre et al., 2017; Müller et al., 2018; Kiel et al., 2017). New modes of working may be required in the future, which might have both beneficial and negative repercussions for employees (Smit et al., 2016). Changes in working environment may result in workplace conflict (Bauer et al., 2015). Interorganizational relationships, as well as the tools and systems utilized by manufacturing firms, might cause standardization challenges. (Nagy, 2019). Some of the common challenges that third-world countries generally face are lack of knowledge and availability and affordability of Industry 4.0.

Concerns about the secure management of private information and data are predicted to expand in the future, according to the development of new technologies (Kovács 2017). It is also worth highlighting the value of privacy-enhancing technologies, which employ technical solutions to protect personal data and privacy (Heurix et al., 2015).

As a result, the evolution of Industry 4.0 has a significant influence on the risk of fragility, increasing the ecosystem's unpredictability (Kovács, 2018). The most major stumbling problem, according to Kiel et al. (2017), is the necessity for technical integration. Because it can be challenging to synchronize several languages, technologies, and processes, successful component, tool, and method integration necessitate the creation of a flexible interface (Zhou et al., 2015).

2.6 PROPOSED MODEL

Industry 4.0 encourages companies to align their business models, which consider information as a CA and are fueled by data collecting, utilization, and sharing opportunities. Plans of action explain an organization's essential objectives, source of advantage, items, and administration portfolio. Maintaining a focused edge in the notion of Industry 4.0 necessitates a transformation plan. The proposed model shown in Figure 2.5 has the following parts.

Ecosystem: The transition from an ad hoc gathering of individuals to a well-organized network is depicted by the business biological system. Another and more diverse type of strategy has emerged around the supposed stages that can function as a commercial hub, thereby connecting supply and demand. Through the communication stream, stages connect individuals in the value chain and communicate data regarding things, administrations, and procedures.

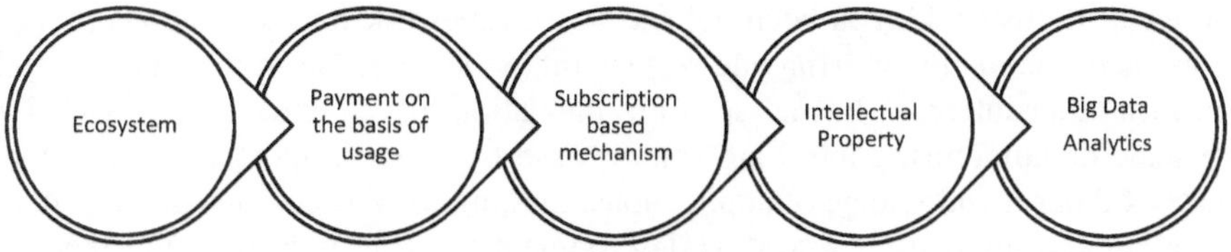

FIGURE 2.5 Proposed model. (*Source:* Authors own.)

- **Payment on the Basis of Usage:** Another strategy for selling things in the manufacturing industry is to use the remuneration per-use strategy. The installation is built not only on the basis of hardware ownership and a fixed cost but also on the basis of use. This strategy provides a provider with options for customer interactions and data collecting via big data analytics (Gebauer et al., 2017; Manimuthu et al., 2019). The methodology allows for increased income per endorser.
- **Subscription-Based Mechanism:** Two additional expanding methods of action are increased flexibility and speedy adaptability. To approach an object in a membership-based company, a customer must pay a membership fee, which is feasible through the IoT. It shows the chance of being able to offer the sit-still time when looking for a machine in a manufacturer's processing plant. It provides an estimated cost for the duration of the agreement, as well as the ability to divide the membership cost (Sung, 2018).
- **Intellectual Property**: IP-based strategies will be embraced by organizations that are active in critical areas of innovation and have the capacity to generate, own, and sell protected invention (Fatorachian and Kazemi, 2018). Even if the assembling organization lacks knowledge and skills, a new strategy based on cutting-edge technology may be constructed on the programming, upkeep, and support of members. Organizations may offer additional services, such as educational classes.
- **Big Data Analytics:** Big data has gradually become a crucial part of economic progress. Joining large data sets opens doors that were previously unthinkable (Gebauer et al., 2017). The foundations on which this methodology is built are investigative skills and programming knowledge. There are a variety of immediate and erroneous tactics. An immediate approach can be shown in the information age through vital items and crowdsourcing. Crowdsourcing may help businesses improve their data-driven decisions for a fraction of the cost. A complex strategy necessitates the use of inside information.

2.7 CONCLUSIONS

Analysts and specialists will be required to develop new ways for assessing and analyzing Industry 4.0's potential as its importance develops. Industry 4.0 is changing not only how products and services are made but also how people live, the environment, means of transportation, and many other facets of how our world works.

As a result, Industry 4.0 has been defined as a framework for autonomous process operations in the manufacturing and service industries. Further, it also has the ability to turn labor-intensive businesses into innovation-driven enterprises, as well as large-scale manufacturing into high-volume assortment development. Furthermore, Industry 4.0 has a wide range of applications, ranging from social insurance to communication entrepreneurs, as well as development to bring industries together and coordination to safeguard them. Industry 4.0 is obviously a collection of technologies that firms may employ to enhance their operations, but the three most critical variables to consider when improving techno-entrepreneurship are price, availability, and awareness of Industry 4.0.

REFERENCES

Adamik, A., Nowicki, M. (2018). Preparedness of companies for digital transformation and creating a competitive advantage in the age of industry 4.0. *Proc Int Conf Bus Excell* 12, 10–24.

Adolph, S., Tisch, M., Metternich, J. (2014). Challenges and approaches to competency development for future production. *Educ Altern* 12, 1001–1010.

Alqaryouti, O., Siyam, N. (2018). Serverless computing and scheduling tasks on cloud: a review. *Am Sci Res J Eng Technol Sci* 40(1), 235–247. http://asrjetsjournal.org/index.php/American_Scientific_Journal/article/view/3913.

Ammar, M., Haleem, A., Javaid, M., Walia, R., Bahl, S. (2021). Improving material quality management and manufacturing organizations system through Industry 4.0 technologies. *Mater Today Proc* 45 (P6) 5089–5096.

Anderson, R. (2008). Implications of the information and knowledge society for education. In J. Voogt, & G. Knezek (Eds.), *International Handbook of Information Technology in Primary and Secondary Education* (p. 5e22). New York: Springer.

Antony J., Kumar M., Madu C. N (2005). Six sigma in small- and medium-sized UK manufacturing enterprises: some empirical observations. *Int J Qual Reliab Manag*, 22(8), 860–874.

Arnold, C., Kiel, D., Voigt, K. I. (2016). How the industrial internet of things changes business models in different manufacturing industries. *Int J Innov Manag*, 20(8), 15-25.

Assante, D., Castro, M., Hamburg, I., Martin, S. (2016). The use of cloud computing in SMEs, *Procedia Comput Sci* 83, 1207–1212. https://doi.org/10.1016/j.procs.2016.04.250.

Baker, N. (1971). *Government and Contractors: The British Treasury and War Supplies*, 1775–1783. London: Athlone Press.

Bauer, W., Hämmerle, M., Schlund, S., Vocke, C. (2015). Transforming to a hyper-connected society and economy – towards an "Industry 4.0.". *Proc Manuf* 3, 417424. https://doi.org/10.1016/j.promfg.2015.07.200.

Blanco-Novoa, Ó., Fernández-Caramés, T. M., Fraga-Lamas, P., Vilar-Montesinos, M. A. (2017). *A Practical Evaluation of Commercial Industrial Augmented Processing in Telecommunications (SINKHROINFO), Kazan 2017.* pp. 1–4, https://doi.org/10.1109/SINKHROINFO.2017.7997540.

Bortolini, M., Ferrari, E., Gamberi, M., Pilati, F., Faccio, M. (2017). Assembly system design in the Industry 4.0 era: a general framework. *IFAC-Papers Online* 50(1), 5700–5705. https://doi.org/10.1016/j.ifacol.2017.08.1121.

Branco, T., Sá-Soares, F., Rivero, A. L. (2017). Key issues for the Successful Adoption of Cloud Computing, *Procedia Comput Sci*, 121115–122, https://doi.org/10.1016/j.procs.2017.11.016.

Brettel, M., Friederichsen, N., Keller, M. and Rosenberg, M. (2014). How virtualization, decentralization and network building change the manufacturing landscape: an industry 4.0 perspective", *Int J Mech Ind Sci Eng*, 8(1), 37-44.

Carlen, J. (2016). *A Brief History of Entrepreneurship: The Pioneers, Profiteers, and Racketeers Who Shaped Our World*. Columbia Business School Publishing.

Cellary, W. (2019). Non-technical challenges of industry 4.0. In: L.M. Camarinha-Matos, H. Afsarmanech, D. Antonelli (Eds), *Collaborative Networks and Digital Transformation, IFIP Advances in Information and Communication Technology*, 568, 3–10. IFIP-Springer. https://doi.org/10.1007/978-3-030-28464-0.

Cheng, B., Zhang, J., Hancke, G. P., Karnouskos, S., Colombo, A.W (2018). Industrial cyber physical systems: realizing cloud-based big data infrastructures, *IEEE Ind Electron Mag* 12(1) 25–35, https://doi.org/10.1109/MIE.2017.2788850.

Choi, K., Chung, S.-H. (2017). Enhanced time-slotted channel hopping scheduling with quick setup time for industrial Internet of Things networks, *Int J DistribSens Netw* 13 (6). https://doi.org/10.1177/1550147717713629.

Cooper, A. (2003). Entrepreneurship: The past, the present, the future. In Zoltan J. A. and D. Audretsch (eds.), *Handbook of Entrepreneurship Research*. Boston: Kluwer.

Day, G. S. (1994). The capabilities of market-driven organizations. *J Market* 58(4), 37–52.

Dombrowski, U., & Wagner, T. (2014). Mental strain as field of action in the 4th industrial revolution. *Proc CIRP*, 17, 100. https://doi.org/10.1016/j.procir.2014.01.077.

Drucker, P. (2015). *Innovation and Entrepreneurship*. Routledge.

Duraković, B. (2011). *Aspektiimplementaciješest sigma sistemakvaliteta u malimisrednjim preduzećima*, Master thesis, Faculty of Mechanical Engineering Sarajevo, 2007, 40.

Erol, S., Jäger, A., Hold, P., Ott, K., Sihn, W. (2016). Tangible industry 4.0: a scenario-based approach to learning for the future of production. *Proc CIRP* 54, 13–18. https://doi.org/10.1016/j.procir.2016.03.162.

Fatorachian, H. and Kazemi, H. (2018). A critical investigation of Industry 4.0 in manufacturing: theoretical operationalization framework, *Produc Plan Control*, 29(8), 633–644.

Fielt, E. (2013). Conceptualising business models: definitions, frameworks and classifications. *J Bus Mod*, 1(1), 85. https://doi.org/10.1177/0149206311406265.

Gebauer, H., Haldimann, M. and Saul, C. J. (2017). Competing in business-to-business sectors through pay-per-use services, *J Serv Manag*, 28(5), 914–935.

Ghicajanu, M. (2015). Innovation and entrepreneurial creativity business during Henry Ford. Annals of the "Constantin Brâncuşi" University of Târgu Jiu, Economy Series, Issue 1, Volume II.

Gilchrist, A. (2016). *Industry 4.0: The Industrial Internet of Things*, Apress, New York, NY.

Goh, T. N., (2011). Six Sigma in industry: some observations after twenty-five years. *Quality*.

Górka, K., Thier, A., Łuszczyk, M. (2020). Consequences of the fourth industrial revolution in social and economic development in the 21st century. In *The Future of Management*, Buła, P., Nogalski, B., (Eds.), Jagiellonian University Press: Kraków, Poland, pp. 60–71.

He, H. et al. (2016). The security challenges in the IoT enabled cyber-physical systems and opportunities for evolutionary computing & other computational intelligence. *2016 IEEE Congress Evolut Comput*. 1015–1021. https://doi.org/10.1109/CEC.2016.7743900.

Heurix, J., Zimmermann, P., Neubauer, T., Fenz, S., 2015. A taxonomy for privacy enhancing technologies. *Comput Secur* 53, 1–17.https://doi.org/10.1016/j.cose.2015.05.00.

Hodgson, G. (2001). *How Economics Forgot History: The Problem of Historical Specificity in Social Science*. London: Routledge.

Holotiuk, F., & Beimborn D. (2017). Critical success factors of digital business strategy. In Leimeister, J. M., Brenner, W. (Eds.): *Proceedings der 13. Interna-tionalenTagung Wirtschaftsinformatik, St. Gallen, S.991-1005.*

Horejší. (2015). Augmented reality system for virtual training of parts assembly. *Proc Eng*, 100, 699–706, https://doi.org/10.1016/j.proeng.2015.01.42276.
Hunt, E. S., Murray, J. 1999. *A History of Business in Medieval Europe*, 1200–1550. Cambridge: Cambridge University Press.
Hunter, M. (2013). A short history of business and entrepreneurial evolution during the 20th century: trends for the new millennium. *Geopol Hist Int Relat* 5, 11–67.
Joško, L., Milković, M., Ines, L. (2019). Entrepreneurship development in the context of postmodern-future of entrepreneurship or adaptation to the development of postmodern society. *49th International Scientific Conference on Economic and Social Development "Building Resilient Society", Zagreb, 13–14 December 2019.*
Jung, K., Choi, S., Kulvatunyou, B., Cho, H., Morris, K. C. (2017). A reference activity model for smart factory design and improvement. *Produc Plan Control* 28(2), 108–122.
Karre, H., Hammer, M., Kleindienst, M., Ramsauer, C. (2017). Transition towards an industry 4.0 state of the LeanLab at Graz University of Technology. *Procedia Manuf* 206–213.
Kiel, D., Arnold, C., Voigt, K.-I., (2017). The influence of the industrial internet of thing son business models of established manufacturing companies – a business level perspective. *Technovation* 68, 4–19.
Kovács, O., (2017). AzIpar 4.0 komplexitása – I. *KözgazdaságiSzle* 64, 823–851.
Kovács, O., (2018). The dark corners of industry 4.0 – grounding economic governance 2.0. *Technol Soc* 55, 140–145. https://doi.org/10.1016/j.techsoc.2018.07.009.
Kumar, S., Suhaib, M., Asjad, M. (2020). Industry 4.0: complex, disruptive, but inevitable. *Manag Prod Eng Rev*, 11, 43–51.
Kurilova-Palisaitiene, J., Lindkvist, L., Sundin, E. (2015). Towards facilitating circular product life-cycle information flow via remanufacturing. *Proc CIRP*, 29, 780–785.
Lachenmaier, J. F., Lasi, H., Kemper, H.-G. (2017). Simulation of production processes involving cyber-physical systems, *Proc CIRP* 62, 577582. https://doi.org/10.1016/j.procir.2016.06.074.
Lasi, H., Fettke, P., Kemper, H. G., Feld, T., Hoffmann, M. (2014). Industry 4.0. *Bus Inform Syst Eng* 6, 239–242. https://doi.org/10.1007/s12599-014-0334-4.
Lee, J., Bagheri, B. and Kao, H. A. (2015). A cyber-physical systems architecture for Industry 4.0-based manufacturing systems, *Manufact Lett*, 3, 18–23.
Lee, J., Kao, H. A., & Yang, S. (2014). Service innovation and smart analytics for industry 4. 0 and big data environment. *Proc CIRP*, 16, 3. https://doi.org/10.1016/j.procir.2014.02.
Levy, F., & Murnane, R. J. (2012). *The New Division of Labour: How Computers Are Creating the Next Job Market.* Oxford: Princeton University Press.
Loebbecke, C., & Picot, A. (2015). Reflections on societal and business model transformation arising from digitization and big data analytics: a research agenda. *J Strategic Inform Syst*, 24(3), 149–157.
Manimuthu, A., Sreedharan, V. R., Rejikumar, G. and Marwaha, D. (2019). A literature review on bitcoin: transformation of crypto currency into a global phenomenon. *IEEE Eng Manag Rev* 47(1), 28–35.
Markussen, S., Røed, K. (2017). The gender gap in entrepreneurship e the role of peer effects. *J Econ Behav Organ* 134 (Supplement C), 356e373.
Mason, C., Brown, R. (2014). Entrepreneurial ecosystems and growth oriented entrepreneurship, Background paper prepared for the workshop organised by the OECD LEED Programme and the Dutch Ministry of Economic Affairs on Entrepreneurial Ecosystems and Growth Oriented Entrepreneurship, Netherlands.
Mehmood, T., Alzoubi, M. (2019). Schumpeterian entrepreneurship theory: evolution and relevance, *Acad Entrepreneur J* 25(4), 1528-2686-25-4-30.
Meyer, M. H., & Tucker, J. M. (2010): Innovating for effectiveness: lessons from design firms, *Res Technol Manag* 53(5), 21–28.

Mourtzis, C., Doukas, M., Bernidaki, D. (2014). Simulation in manufacturing: review and challenges, *Procedia CIRP* 25, 213–229, https://doi.org/10.1016/j.procir.2014.10.032.

Mourtzis, D., Zogopoulos, V., Vlachou, E. (2017). Augmented reality application to support remote maintenance as a service in the robotics industry, *Proc CIRP* 63, 46–51. https://doi.org/10.1016/j.procir.2017.03.154.

Müller, J. M., Buliga, O., Voigt, K.-I. (2018). Fortune favors the prepared: how SMEs approach business model innovations in industry 4.0. *Technol Forecast Soc Chang*, 132, 2–17. https://doi.org/10.1016/j.techfore.2017.12.019.

Nagy. (2019). Azipar 4.0 fogalmaéskritikuskérdései – vállalatiinterjúk alapján. *Vezetéstudomány* 50, 14–26. https://doi.org/https://doi.org/10.14267/VEZTUD.2019.01.02.

Negahban, A., Smith, J. S. (2014). Simulation for manufacturing system design and operation: literature review and analysis, *J Manuf Syst* 33(2), 241–261, https://doi.org/10.1016/j.jmsy.2013.12.007.

Ning, H., & Liu, H. (2015). Cyber-physical-social-thinking space based science and technology framework for the Internet of things. *Sci China Inform Sci* 58, 1–19. https://doi.org/10.1007/s11432-014-5209-2.

Palmarini, R., Erkoyuncu, J. A., Roy, R., Torabmostaedi, H. (2018). A systematic review of augmented reality application in maintenance. *Rob Comput Integr Manuf* 49, 215–228, https://doi.org/10.1016/j.rcim.2017.06.002.

Parrott, D. (2012). *The Business of War.* Cambridge: Cambridge University Press. https://doi.org/10.1017/CBO9781139023337.

Pedersen, M. R. Nalpantidis, L., Andersen, R. S., Schou, C., Bøgh, S., Krüger, V., Madsen, O. (2016). Robot skills for manufacturing: from concept to industrial deployment, *Rob Comput Integr Manuf* 37, 282–291. https://doi.org/10.1016/j.rcim.2015.04.002.

Perez, C. (2009). Technological revolutions and techno-economic paradigms. *TOC/TUT Working Paper*, 20, 6.

Peteraf, M. A. (1993). The cornerstones of competitive advantage: a resource-based view. *Strat Manag J* 14(3), 179–191.

Petrelli, D. (2017). Industry 4.0: is it time for interaction design craftsmanship?. *Design J*, 20(S1), S2735–S2745.

Prakash, C., Saini, R., & Sharma, R. (2021). R. Sharma, R. Saini & C. Prakash, V. Prashad, *Role of Internet of Things (IoT) in Sustaining Disruptive Businesses* (1st ed.). New York: Nova Science Publishers.

Qi, Q., Tao. F. (2018). Digital twin and big data towards smart manufacturing and industry 4.0:360 degree comparison, *IEEE Access* 6, 3585–3593. https://doi.org/10.1109/ACCESS.2018.2793265.

Rakshit, P. & Sharma, R. (2020). Building healthcare 4.0 with smart workforce. In *Proceedings of the International Conference on Research in Management & Technovation*, PTI. Vol 24. pp. 91–94. 10.15439/2020KM5.

Rana, G., & Sharma, R. (2017). Organizational Culture as a Moderator of the Human Capital Creation-Effectiveness. *Glob HRM Rev*, 7(5), 31–37.

Rana, G., & Sharma, R. (2021a). Employer branding: attracting and retaining employees for sustainable development in disruptive economy. *World Rev Sci Technol Sustain Dev*, 17(4), 319. https://doi.org/10.1504/wrstsd.2021.117839.

Rana, G., & Sharma, R. (2021b). Impact of competency mapping on talent management for improving business performance. *Asian J Multidimen Res*, 10(3), 310–315. https://doi.org/10.5958/2278-4853.2021.00182.8.

Rana, G., Sharma, R., Goel, A. K. (2019). Unraveling the power of talent analytics: implications for enhancing business performance. In: Rajagopal, B. R. (eds.) *Business Governance and Society*. Palgrave Macmillan, Cham. https://doi.org/10.1007/978-3-319-94613-9_3.

Renjen, P. (2021). Deloit report. Retrieved from https://www2.deloitte.com/content/dam/insights/us/articles/us32959-industry-4-0/DI_Industry4.0.

Rentzos, L., Papanastasiou, S., Papakostas, N., Chryssolouris, G. (2013). Augmented reality for human-based assembly: using product and process semantics. *IFAC Proc Vol*, 46(15), 98–101. https://doi.org/10.3182/20130811-5-US-2037.00053.

Rodic, B. (2017). Industry 4.0 and the new simulation modelling paradigm, *Organizacija* 50(3), 193–207, https://doi.org/10.1515/orga-2017-0017.

Sadiku, M. N. O., Wang, Y., Cui, S., Musa, S. M. (2017). Industrial internet of things. *Int J Adv Sci Res Eng* 3(11), 1–5, https://doi.org/10.7324/IJASRE.2017.32538.

Salkin, C., Oner, M., Ustundag, A., Cevikcan E. (2018). *A Conceptual Framework for Industry 4.0, In Industry 4.0: Managing the Digital Transformation, Springer Series in Advanced Manufacturing*, Springer, Cham, 3–23. 10.1007/978-3-319-57870-5.

Schweikart, L. & Doti, L. P. (2010). *American Entrepreneur*, New York: AMACON, 309.

Sergey, M., Nikolay, S., Sergey, E. (2019). Cyber security concept for internet of everything (IoE). *2017 Systems of Signal Synchronization, Generating and Processing in Telecommunications (SINKHROINFO), Kazan 2017. 1–4.* https://doi.org/10.1109/SINKHROINFO.2017.7997540.

Sezer, O. B., Dogdu, E., Ozbayoglu, A. M. (2018). Context-aware computing learning, and big data in internet of things: a survey, *IEEE Internet Things J* 5(1), 1–27, https://doi.org/10.1109/JIOT.2017.2773600

Shamim, S., Cang, S., Yu, H., Li, Y. (2016). Management approaches for industry 4.0: a human resource management perspective.

Shao, G., Shin, S. J., Jain, S. (2014). Data analytics using simulation for smart manufacturing, *Proceedings of the Winter Simulation Conference 2014, Savanah*, pp. 2192–2203, https://doi.org/10.1109/WSC.2014.7020063.

Sharma, R., Singh, S., Rana, G. (2019). Employer branding analytics and retention strategies for sustainable growth of organizations. In: *Understanding the Role of Business Analytics*, pp. 189–205; Springer Singapore. https://doi.org/10.1007/978-981-13-1334-9_10.

Sharma, R., & Rana, G. (2021). An overview of entrepreneurship development programs in India. *Int J Environ Policy Decision Mak*, 3(1), 16. https://doi.org/10.1504/ijepdm.2021.119511.

Sharma, R., Rana, G., Agarwal, S., (2021a). Techno innovative tools for employer branding in industry 4.0. In: G. Rana, S. Agarwal and R. Sharma, ed., *Employer Branding for Competitive Advantage*, 1st ed. Boca Raton, FL: CRC Press. https://doi.org/10.4324/9781003127826-11.

Sharma, R., Saini, R., Prakash, C., Prasad, V. (2021b). *Internet of Things and Businesses in a Disruptive Economy*, 1st edn., New York: Nova Science Publishers.

Sharma, R., Saini, A. K., Rana, G. (2021c). Big data analytics and businesses in industry 4.0. *Design Eng* 2021(02), 238–252.

Sharma, Manu, Luthra, Sunil, Joshi, Sudhanshu and Kumar, Anil (2021). Implementing challenges of artificial intelligence: evidence from public manufacturing sector of an emerging economy. Government Information Quarterly. ISSN 0740-624X.

Singh R., Anita G., Capoor S., Rana G., Sharma R., Agarwal S. (2019). Internet of things enabled robot based smart room automation and localization system. In: Balas V., Solanki V., Kumar R., Khari M. (eds) *Internet of Things and Big Data Analytics for Smart Generation. Intelligent Systems Reference Library*, vol. 154. Springer, Cham. https://doi.org/10.1007/978-3-030-04203-5_6.

Sipsas, K., Alexopoulos, K., Xanthakis, V. and Chryssolouris, G. (2016). Collaborative maintenance in flow-line manufacturing environments: an Industry 4.0 approach. *Proc CIRP*, 55, 236–241.

Smit, J., Kreutzer, S., Moeller, C. & Carlberg, M. (2016). Industry 4.0. Brussels.

Sung, T. K. (2018). Industry 4.0: a Korea perspective. *Technol Forecast Soc Change*, 132, 40–45.

Tao, F., Cheng, J., Qi, Q., Zhang, M., Zhang, H., Sui, F. (2018). Digital Twin-driven product design, manufacturing and service with big data, *Int J Adv Manuf Technol* 94 (9–12) 3563–3576, https://doi.org/10.1007/s00170-017-0233-1.

Thomas A., Lewis G. (2007). Developing an SME-based integrated TPM-Six Sigma strategy. *Int J Six Sigma Compet Adv*, 3, 228–247.

Tsuchiya, A., Fraile, F., Koshijima, I., Ortiz, A., Poler, R. (2018). Software defined networking firewall for industry 4.0 manufacturing systems. *J Ind Eng Manag* 11(2), 318–333. https://doi.org/10.3926/jiem.2534.

Vyas, D., & Nijholt, A. (2016). From Mundane to smart: exploring interactions with 'smart' design objects. *Int J Mob Human Comp Interact*, 8(1), 59. https://doi.org/10.4018/IJMHCI.2016010103.

Xu, X. (2012). From cloud computing to cloud manufacturing, *Rob Comput IntegrManuf*, 28 (1) 75–86, https://doi.org/10.1016/j.rcim.2011.07.002.

Yin, S., Kaynak, O. (2015). Big data for modern industry: challenges and trends [point of view]. *Proc IEEE* 103 (2) 143–146, https://doi.org/10.1109/JPROC.2015.2388958.

Zhou, K., Liu, T., Zhou, L. (2015). Industry 4.0: towards future industrial opportunities and challenges. In *12th International Conference on Fuzzy Systems and Knowledge Discovery (FSKD). Zhangjiajie, China*, pp. 2147–2152

Zúñiga, E. R., Moris, M. U., Syberfeldt, A. (2017). Integrating simulation-based optimization, lean, and the concepts of industry 4.0. *2017 Winter Simulation Conference (WSC), Las Vegas, NV 2017.* 3828–3839. https://doi.org/10.1109/WSC.2017.8248094.

3 Individual and Contextual Factors in Start-Ups Financing Decision

Insights from Moroccan Technological Start-Ups

Widad Ahl Maatallah
Tangier International Higher Institute of Tourism (ISITT)

Mostapha El Idrissi
Cadi Ayyad University

CONTENTS

DOI: 10.1201/9781003217084-3

3.1 INTRODUCTION

Start-ups and entrepreneurial firms are the cornerstone in every economy as they fuel economic development, support growth of industries and markets, and promote employment (Bellavitis et al. 2017; Megginson 2004; Rasmussen et al. Sørheim 2012). In reaching success and growth goals, start-ups depend on financing sources. Due to their financial shortcomings, trading history, limited collateral and insufficient information, start-ups often encounter inevitable difficulties in accessing and getting funding (Brown, Mawson, et al. Rowe 2018; Hussain, Salia, et al. Karim 2018). In response, entrepreneurs often handle this need by resorting to external financing sources (Block et al. 2018).

In recent years, a set of new and suitable sources of financing have emerged to improve entrepreneurial finance ecosystem and compensate the inadequacy of traditional sources for young firms. These alternatives include a variety of actors, accelerators, and incubators as important guides, proof-of-concept centers and universities as source of nascent entrepreneurs and innovative ideas, crowdfunding platforms, and other financial instruments as funding players (Bellavitis et al. 2017). In addition, traditionally, venture capital and business angel financing are considered as important sources of financing for start-ups that have difficulty in obtaining bank and debt financing (Block et al. 2018).

Nascent entrepreneurs, specifically in emerging economies, are underinformed about the existence of such new financing players, and then they prefer simple and familiar sources. Bellavitis et al. (2017) argue that the institutional voids in emerging markets are the cause of the accessing difficulties to external financing. Consequently, the development of venture capital and debt markets in such economies was negatively affected by these institutional problems (Bruton, Fried, et Manigart 2005).

Existing literature has devoted an effort to understand the financial problems of nascent firms through listing and explaining key elements behind financial difficulties (e.g., asymmetric information and agency problems), facing those type of firms in accessing to external funding (Block et al. 2018; Hall et Lerner 2010). Despite the interesting contributions of previous researches, we cannot completely understand the financing problems of young firms by relying only on old theories (e.g., agency theory or pecking order theory). Moreover, the entrepreneurial finance's landscape has changed and some innovation on the analytical approach is required (e.g., human approach). Since the strategic decisions and particularly the financing decisions are held by entrepreneurs and start-ups' owners, it is interesting to investigate the financing decision process by integrating the perception of entrepreneurs and exploring factors that shape this perception.

Reviewing the demand-side entrepreneurial finance landscape, we came up with the fact that start-ups' financing is a multifactorial decision-making process. Hence, scholars argue that entrepreneur's human and social capital (Heuven et al. Groen 2012; Seghers, Manigart, et al. Vanacker 2012; Shepherd, Williams, et al. Patzelt 2015) coupled with contextual environment (Almobaireek, Alshumaimeri, et al. Manolova 2017; Mac an Bhaird et al. Lucey 2014; Vanacker, Heughebaert, et al. Manigart 2014) could explain the financing decision-making.

In this chapter, we proceed by exploring and identifying the core elements of individual and contextual factors that shape the perception of Moroccan entrepreneurs about the external financing alternatives. We believe that investigating new settings will reveal interesting conclusions that enlighten the inconsistency of previous empirical findings. For a second time, we believe that the contextual and individual factors can act differently on the financing decision-making process; hence, we investigated the manner in which they influenced and classified them as impeding factors and incentive factors.

We conducted multiple case studies on 12 Moroccan technological start-ups to unveil how individual and contextual factors influence the start-ups' financing decision choice. Following a qualitative approach, we are able to grasp how the entrepreneur's perception of financing alternatives is shaped and recognize the elements underlying it.

In the first section of this chapter, we introduce the topic of start-ups' financing choices. The second section will be dedicated to individual and contextual factors that shape the entrepreneurs' perception of financing choices. In the third section, we present the method used in this chapter. After which, we'll expose the main results. Discussion and conclusion are provided in the final section.

3.2 START-UPS AND FINANCIAL DIFFICULTIES

3.2.1 Start-Ups' Definition

Due to differences in conventions across countries, defining start-ups' firms is complicated. For many authors, start-ups firm is a young entity (less than 5 years of existence) that presents an innovative idea and demonstrates a high growth potential, or at least, one of those criteria. Hall (2002) considers that start-ups are generally technological firms, which need financing resources in a permanent way. Indeed, nowadays, new and young firms are more prone to exploit new technologies and knowledge (Braunerhjelm 2013). In this sense, several public institutions attribute the technological aspect to start-up; similarly, Moroccan institutions[1] also define business start-ups as technological new firms. The programs they develop to accompany newborn firms include technology that is used as an important criteria in the selection of potential beneficiary firms. In this chapter, we follow Luger and Koo' (2005) who define start-ups as new, active, and independent enterprises (Figure 3.1). "New" describes the newness of the creation idea, the entrepreneurial characteristics, and the unsophisticated ways of making business and management decisions, which denotes the main discriminator for many studies (Chin 2020; Ren et al. 2020). "Active" refers to firms that have been in business for at least 1 year since creation and currently have minimum one employee. "Independents" start-ups are those that are neither dependent on other firms or groups nor are a spin-off from other companies (Witte et al. 2018).

[1] For example, La chambre française de commerce et d'industrie, CFCIM, which is a private association, under Moroccan law, whose mission is to promote economic relations between France and Morocco, encourage the establishment and development of companies in the United Kingdom, and develop the international potential of Moroccan companies, defines start-ups as young businesses based on technology or internet.

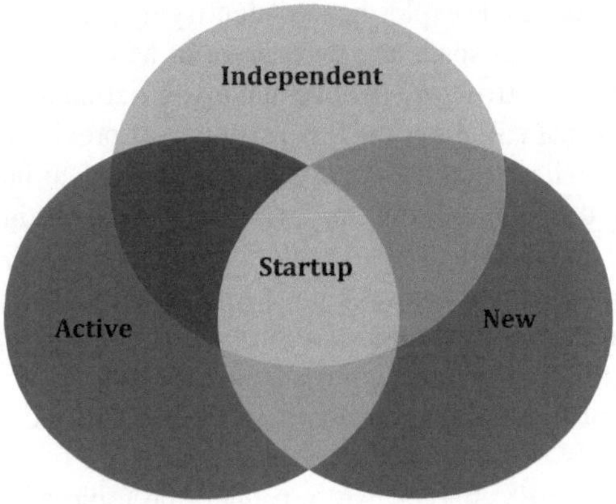

FIGURE 3.1 Determinant's characteristics of start-up firms.

According to Oviatt and McDougall (1995), firms that internationalize need a stage of domestic maturation, which means that start-ups are usually not international firm. Recently, owing to technological development, distances have become virtually closer or even disappeared, hence, nothing has become more international than a technological start-up. Another main characteristic of start-ups is that the firm's strategies are dependent on the founder, who executes a dominant influence on their business (Julien 1990; Marchesnay 1991).

3.2.2 Start-Ups' Financial Difficulties

Regardless of how good or convincing the initial idea is, only half of the start-ups can survive for more than 5 years, and only the most promising can get early professional investment. Indeed, even with extensive financing, more than 75% of venture-backed start-ups fail or remain with a minor presence (Picken 2017). An overview of literature suggests that the main cause of a start-up firms' failure is financing deficit (Adomako, Danso, et Ofori Damoah 2016; Atherton 2012; Pellegrino, Piva, et al. Vivarelli 2011). Due to insufficient funds to finance their business activity, the firm's owner may need to look for a suitable external financing (Atherton 2012; Gompers and Lerner, 2010). For instance, start-ups may not have an effective ability to establish adequate equity to fuel their growth and develop their projects. Hence, these firms rely more on loans and external support (Brown et al. 2018).

Since start-ups are usually created to concretize innovative new ideas, they are considered as firms that are characterized by a large number of intangible assets, they cannot expect positive earnings on the first years of creation, and have uncertain prospects, in addition, are unlikely to receive bank loans or other debt financing (Block et al. 2018).

For some authors, it is familiar that low-wealth entrepreneurs are less likely to attain some forms of external financing (Brush, Edelman, et al. Manolova 2012; Frid et al. 2016). Many models of ownership and financing choice are built upon the inability of investors to check whether certain actions have been taken or certain outcomes have occurred, but regarding the verification of the characteristics of firm's selection, they found that angels are interested in management's ability and market potential, the state of the entrepreneurial team, marketing, and finance. Start-up firms are probably going to confront precisely these kinds of issues, making outside financing expensive or hard to obtain (Gompers et Lerner 2010).

3.3 FACTORS SHAPING THE PERCEPTION ABOUT FINANCIAL CHOICES

In an examination of the start-up financing framework, we notice that the financial decision making process is mainly influenced by individual aspects of the entrepreneur (Heuven et al. Groen 2012; Hussain et al. 2018; Nouira, Klofsten, et al. Dahlstrand 2005; Seghers et al. 2012) as well as the contextual environment (Almobaireek et al. 2017; Mac an Bhaird et Lucey 2014; Vanacker et al. 2014).

3.3.1 Individual Factors

Education may be considered unimportant by finance providers; however, in fact, it seems to be an important demand-side factor of venture capital financing. Higher education can give entrepreneurs the tools they may need to negotiate effectively with potential finance providers (Coleman, Cotei, et al. Farhat 2016; Hogan et al. Hutson 2005). More specifically, researchers point out that having a formal education in domains related to finance and business can increase the likelihood of understanding and using financial alternatives by entrepreneurs (Hussain et al. 2018; Seghers et al. 2012). This specific education can mitigate information asymmetry, reduce monitoring costs, and improve capital flows (Hussain et al. 2018), which enable entrepreneurs to make adequate strategic investments and deal with complex financial decisions (Matthews et al. 2013). Similarly, Seghers et al. (2012) show that business education is correlated with a broader knowledge of finance alternatives. In contrast, Cassar (2004) found that decision makers' characteristics (education, experience, and gender) have no significant relationship with start-up financing.

Obtaining funding is a complex and social process; as a result, connections and networks of entrepreneurs are key factors to understand and access external funding (Rasmussen et Sørheim 2012). Consequently, in addition to education, entrepreneurs can also acquire information about financing options through their social capital (Hsu 2007; Seghers et al. 2012).

Social capital researchers focus on the external social ties of individuals as a resource that may explain the individual's success as they can rely on their networks, contacts, and connections to seize opportunities and access information in the external environment (Adler et Kwon 2002; De Carolis et Saparito 2006; Granovetter 1985). Indeed, researchers point out that the act of entrepreneurship is a consequence of social interaction. Accordingly, the social networks where the entrepreneurs are embedded have an important role in recognizing and creating opportunities (De Carolis et Saparito 2006; McKeever, Anderson, et Jack 2014).

Traditionally, start-ups' owners often suffer from limited information about available financing choices (Van Auken 2005; Seghers et al. 2012). This lack of information may be filled by the social capital of the entrepreneur, since the network can act as a bridge to identify and access financial resources (Heuven et Groen 2012; Rasmussen et Sørheim 2012).

Seghers et al. (2012) claim theoretically and confirm empirically that entrepreneurs with network ties in the financial community have greater knowledge about common finance alternatives, which may reduce information problems experienced by entrepreneurs. Moreover, Shepherd et al. (2015) expect that entrepreneurs who are virtually connected may find new funding alternatives (e.g., crowdfunding) more attractive than traditional ones. More recently, Brown et al. (2018) demonstrate that networks and social capital assume a basic role in the equity crowdfunding measure for new firms.

3.3.2 Contextual Factors

Reviewing the literature on start-ups and institutions, arrive at the fact that institutions (formal and informal) are important in driving the behavior of firms and shaping the strategic choices of nascent entrepreneurs (Ahlstrom et al. Bruton 2002; Almobaireek et al. 2017; Ekanem, Owen, et al. Cardoso 2019).

Informal institutions include constraints embodied in traditions, culture, and values, which are developed over time through social interactions among different participants and shape their opinions on what actions are convenient and conceivable (Bruton, Ahlstrom, et al. Singh 2002). In fact, Almobaireek et al. (2017) point out that access to capital within the Islamic countries is solidly impacted by Islamic rules of financing, the most crucial of which is the strict prohibition on giving or receiving any fixed rate of return on financial transitions, more popularly referred as the ban on interest. In addition, Bruton et al. (2002) highlight not rapidly scalable or limited start-ups but the locally focused ones and those that are non-technology-oriented. These start-ups will be influenced by the cultural/cognitive institutions in shaping their choices. In contrast, businesses focusing on broader markets and rapidly changing technology-oriented start-ups will gain from the expertise of worldwide-oriented investors.

On the contrary, formal institutions encompass institutional rules, laws, and regulations implemented by governments to adjust and promote the entrepreneurial ecosystem. Governments around the world take into account the new instruments of financing to promote the creation and growth of start-ups, those new players are instruments to ease the financing limits and difficulties of entrepreneurial firms (Block et al. 2018). The rapid growth of the new instruments of financing is intimately related to the financial and economic crisis of 2008 (Brown et al. 2018). As a response, policymakers intensified the regulation of financial institutions, especially banks. To comply with these intensified regulations, various risk measures had to be introduced by banks that make start-ups' financing more complex and expensive. As a consequence, chances to obtain bank financing by start-ups and small firms with uncertain and risky business models are extremely low (Block et al. 2018).

3.4 METHOD

Traditionally, research on entrepreneurship relies primarily on a quantitative approach (Suddaby, Bruton, et Si 2015). Nevertheless, some scholars advocate the importance of qualitative research in exploring new and complex phenomenon as entrepreneurial finance (St-Pierre et Fadil 2016). Therefore, we consider that qualitative research is a suitable approach to grasp the perception of entrepreneurs and the factors that influence their choices *vis-à-vis* the available financial alternatives since it may help us to recognize entrepreneurs' mindsets and behavior during the decision-making process, which would be challenging to uncover with other research strategies. These mindsets and actions were explored in their natural settings, which help us explain and better understand the phenomenon under study (Miles, Huberman, et Saldaña 2014).

We follow a case study design, which provides us a better understanding of complex social phenomenon and investigate organizational/managerial behavior as decision-making process within start-ups (Yin 2009). We suppose that the case study approach will serve as an illustration of factors that influence the decision-making process in a particular context–Morocco. In order to guarantee the replication (Eisenhardt 1989), we used multiple case studies method to establish an analytical generalization over statistical one (Yin 2009). Indeed, relying on various case studies bring different insights to explore the same subject, which enable us to have robust findings (Eisenhardt et Graebner 2007; Yin 2009).

3.4.1 Data Collection

Semistructured interviews with Moroccan entrepreneurs are the main tools of data collection. Interviews will maximize our chance to collect extensive empirical information, particularly when the subject under study is occasional and infrequent (Eisenhardt et Graebner 2007) as it was the decision of funding for start-ups. In addition, semistructured interviews give respondents greater freedom to talk about topics that seem more important to them (Denzin et Lincoln 2018). Yet, questions included in the interview were predetermined to make sure that it is focused on the research question (Gioia, Corley, et Hamilton 2013).

For triangulation objective, report and document examination were performed. To reinforce our analysis, we also used secondary data from different sources, such as websites of start-ups, reports of public and private institutions, and specialized press reviews. In our opinion, the secondary data will be very relevant and complementary to those collected from the interviews. Secondary data are likely to be more objective and provided us with interesting information about the start-ups being studied. Indeed, it helped us to create a history of each start-up, their starting point (birth) and their development. In addition, it improved our understanding of the overall context of entrepreneurship and the development of financing alternatives in Morocco.

3.4.2 Case Selection and Description

To better understand the perception of entrepreneurs on the themes analyzed, we studied 12 cases of start-ups in Morocco that were chosen heterogeneously. In accordance with Eisenhardt's (1989) recommendations, we pursue a theoretical sampling to select cases that are consistent with the theoretical question and useful for replication. Hence, we have selected start-ups that are in their early stage (all of them are born after 2015) and that use a variety of financial alternatives. Table 3.1 presents

TABLE 3.1
Characteristics of Start-Ups

Start-Up Code	Respondent	Nature of Business	Financing Options Used
S1	Co-founder	Web and mobile development	• Institutional funds • Personal funds • Private incubation
S2	Co-founder	Mobile development	• Institutional funds • Personal funds • Business angel
S3	Co-founders	Design and 3D printing	• Institutional funds
S4	Founder	Web and mobile development	• Institutional funds • Personal funds
S5	Founder	Flooring industry	• Institutional funds • Public Incubation • Proof-of-concept funding • Bank loans
S6	Co-founder	Education	• Personal funds • Venture Capital
S7	Founder	Education	• Personal funds • Institutional funds
S8	Founder	New mobile app	• Institutional funds • Personal Funds
S9	Founder	Software creation	• Personal funds • Institutional funds • Venture capital
S10	Founder	Digital security	• Personal funds • Institutional funds
S11	Founder	Communication and augmented reality	• Institutional funds • Personal funds
S12	Founder	Software creation	• Personal funds • Bank loans

the main characteristics of the start-ups studied and illustrates the heterogeneity obtained in its constitution, especially in terms of financial funding.

3.4.3 Data Analysis

After harmonizing all the data collected from the field, we drew a chronological history of each case to create our own overall understanding of the study and the reality as presented by the respondents (Langley 1999).

We then coded the data to clarify the interpretation and deepen the analysis. The process of data analysis was exploratory, where emerging interpretations were continually reinterpreted as coding progressed (Corbin et Strauss 2007). During our analysis, we followed an abductive process, and hence, deductive and inductive coding methods were performed successively (Miles et al. 2014).

First, we proceed with deductive coding by identifying some first-order codes that are rooted in the literature. While we proceed with the analysis, other codes inductively emerged from the field. As we continued analyzing the data, we understood more about our case studies, and some first-order codes were deleted when they were repetitive or overlapped with others. Subsequently, we started to seek similarities and differences to organize the first-order codes into second-order ones. After several rounds of the coding process, we grouped similar second-order codes into aggregate categories (Gioia et al. 2013) as shown in Figure 3.2.

We relied on Nvivo 12 software to analyze the qualitative data of our research. Compared to manual analysis, Nvivo 12 software allows us to store, edit, and organize large data through a meaningful coding process (Ghauri et Firth 2009; Miles et al. 2014). The data analysis was conducted in the original language (French); therefore, the results and supporting quotes used in this research were translated into English.

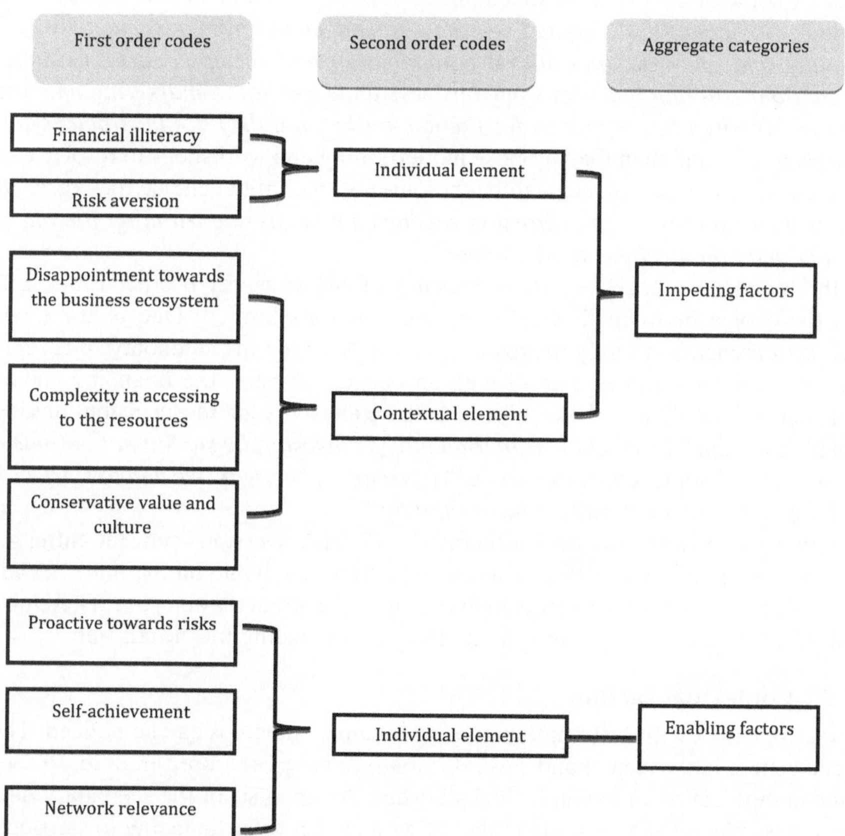

FIGURE 3.2 Nodes and categories of thematic analysis.

3.5 FINDINGS

Start-ups case studies show certain heterogeneity in the use of external financing options, which rely on two types of factors, the incentive factors and the impeding factors, as noted in our thematic analysis.

3.5.1 Impeding Factors

In their financing choice process, entrepreneurs face factors that constrain and limit their choice. Our investigation shows that impeding factors can be linked to the entrepreneur's individual aspect as well as to the particularities of the start-up's context.

3.5.1.1 Individual Factors

Following discussions with entrepreneurs we met, we noticed that their professional profiles are mostly technical with engineering education. As their projects are conducted, they become familiar with the entrepreneurial environment and gradually discover the inadequacies in the financial culture. As a result, they will not recognize and pursue some of the available financing options. Aware of this insufficiency, they argue that: "*someone who is going to create a startup comes from a technical world, it's usually someone who does not have financial or/and entrepreneurial culture,*" or "*even if our training includes some business units like sales, marketing and law which are supposed to prepare us for the job market, but they are insufficient and very superficial,*" and then they propose more training and workshops to resolve this insufficiency, which they perceive to be the cause of the limited choice they face: "*It would be important to set up awareness-raising campaigns and trainings for young graduates and non-graduates full of ideas.*"

In the founders' speech, a certain tendency of risk passivity has been raised. It makes the choice between financing options somehow limited. One of the interviewed entrepreneurs strongly expressed the fear of losing his autonomy in case he directly or indirectly shares control with an outside investor, the financing choice he will pursue should not make him lose the exclusivity of the decision-making within his start-up, "*if I would consider an outside investor, he should not become a decision-maker,*" nor threaten its survival "*to secure the activity, we had to start with something that does not require a large funding.*"

The two parameters "financial illiteracy" and "risk aversion" prevent entrepreneurs from comparing the different financing options available on the market, thus limiting accessibility to more adequate financing sources that they can use. Therefore, we consider that these factors impede the choice of financing for the start-up.

3.5.1.2 Contextual Factors

In Morocco, a goodwill to strengthen the ecosystem of start-ups can be noticed. The financial sphere has widened and several initiatives have been born in order to support the multiplication of technological start-ups. An analysis of the secondary data makes it possible to list a certain number of entities that offer financing to start-ups, and the most popular are as follows. The first is "Maroc Numeric Fund" (MNF), the first fund dedicated exclusively for start-ups, which intervenes directly or indirectly through participations in the capital of innovative companies in the creation and post-creation phases. The second one is "Innov Invest" fund, a financing mechanism of the La Caisse Centrale de Garantie (La CCG) (Central Guarantee Fund) sponsored by the World Bank with a total price of $50 million and the third is *Réseau Entreprendre Maroc* (REM), which provides free interest loans, about $5,000 per partner over 5 years, to project holders with high growth potential. In addition to these three examples, there are about 12 structures that offer a variety of financing instruments for start-ups and Small Medium sized Enterprises (co-financing, financing, no repayment loan, loan, guarantee, and grant).

Despite all the efforts made by public and private institutions toward a proliferation of financing instruments, the entrepreneurs' perception that they have developed toward the context in which they interact is problematic. Practically, it turns out that, in general, the context is not in favor of a suitable development of the Moroccan technological start-up ecosystem. Entrepreneurs have developed feelings of anger, disappointment, and dissatisfaction toward the funding opportunities their ecosystem allows them, they expressed that: "*our wish is to be financed by a Business Angel, but we think that we fell to the wrong ecosystem, if we were in another context it would have been different*," "*I come back to Réseau Entreprendre Maroc, the basic idea that they sold us is the network, but when we are there, there is no network, because the guy who is the mentor is busy, the same for the other entrepreneurs*." This disappointment, which prevents entrepreneurs from trusting the offers that are basically addressed to them, is favored by a kind of difficulty in accessing financing. "*About associations and financing institutions, I have seen "la CGEM,"*[2] but I was sent to do some boring, because I'm not big enough for them," "we consulted Maroc Numeric Fund for fundraisers, about the "Bayti" project, but we were not selected." Moroccan entrepreneurs are overwhelmed by this general impression of a not adapted context with the nature of the start-up.

The entrepreneur's 'cultural background can also influence their financing decisions and preferences. In a country like Morocco, Muslim culture is very influential and integrates the entrepreneurs' mental schemes and guides their decisions. One of the interviewed start-ups founders reveals that the origin of the external money he would use or/and search matters, he said:

> In my financial decisions, I involve ethics, yes, I have values in business, someone with different or opposed ones couldn't be my partner, I will not accept, if a Business Angel invests in the vineyard and therefore alcohol, then make a financing offer to me, I have reasons that will not allow me to accept him being my funder.

3.5.2 Incentive Factors

In contrast to impeding factors, enabling factors allow the entrepreneur to identify and understand the financing options available in the markets and thus make the appropriate choice of financing for his start-up. According to our analysis, enabling factors are mainly individual aspects embodied in the entrepreneurs. As a result, we did not recognize any positive perception regarding the contextual factors by the respondents.

---

[2] Confédération générale des Entreprises Marocaines: General confederation of Moroccan firms.

3.5.2.1 Individual Aspects

The subjective decision-making process leads the entrepreneurs to develop some compromise with the risk unconsciously. In the investigated start-ups, we notice that the interviewees highlight an interaction between the risk perception, its acceptance, and the potentials measure that can alleviate it. Proactivity toward the risk places the founding entrepreneur in a situation where the perceived risk of using a financing option is reduced following risk-sharing compromises: "*If an investor is ready to bet $ 1 million on a project that may be going to bring him back more in 5 or 6 years, I will agree to see and discuss the proposal with him*," or following a risk negotiation:

> There is no problem of sharing control, at a stage limited, because we can if you want to allocate a small quota to the equity that we will give so that we remain the decision makers and we guide ourselves the company, we cannot exceed 15% of margin that we can allot".

This allows the entrepreneur to integrate financing options from the list of choices that he or she sees as possible.

In addition, when in business start-ups, the founder is carried away by the success, self-realization, and his personal objectives constitute the strategic vision of his start-up. For this reason, the entrepreneur seeks the most compatible sources of financing that will allow him realize these objectives. For example, the founder "S10 start-up" affirms that "*abroad, capital venture is a key alternative to finance successful startups. Moving abroad and changing the market is a huge investment, but for me it worth.*" It is therefore concluded that self-realization drives entrepreneurs to seek more funding opportunities. Start-up's finance decisions and choices do not follow the assumptions of traditional financial theory, which is related to profit maximization and costs minimization, personal factors, matter more.

The third individual factor that we qualify as an incentive is the entrepreneur's personal and professional network. In our study, we noticed that most entrepreneurs seek to strengthen their networks, and according to their remarks, obtaining funding is dependent on the network, even for financing alternatives that are difficult to obtain, in terms of requirements and guarantees, such as bank financing, and the network can facilitate access to bank financing and help achieve the desired amount in their perceptions:

> A start-up in general has no guarantee, and the banker imposes guarantees, a start-up that is what he has as a guarantee? Apart from if he is a son of "in general" the sons of do not have this problem there, they already have funding and a network to divert the guarantees

It is a nonfinancial resource that facilitates access to financial resources, serves as a source of information and advice,

> I would have liked to have contact with the financier even if he does not take my project, to tell me the reason for which the project was rejected or why it is not good, at least he will teach me reasoning otherwise

and pushes the entrepreneur to accept risky financing options where there is sharing control and the decision:

> I'm transparent, the day there will be someone who wants to invest, that's ok no problem, everything is clear everything is signed, sometimes I need someone for me 'assist in decisions and advise me, because the advantage of the investor too is that he will open a network, if I have someone who will invest $ 10.0000 and he will make me a turnover of $ 1,000,000 because it has a solid network, there is no problem I sign

The network allows the start-up to build relationships with trusted stakeholders that inspire entrepreneurs to seek more financing and then agree to use financing options through the network potential that can provide.

3.6 DISCUSSION AND CONCLUSION

The process of obtaining adequate and appropriate finance is filled with difficulties, particularly for start-ups (Seghers et al. 2012), especially when financing decision is a key process for their survival and growth. In fact, some researchers advocate that new ventures' financing decisions are mainly influenced by individual and contextual factors (Mac an Bhaird et Lucey 2014; Zahra, Wright, et Abdelgawad 2014).

In this chapter, we strived to improve our comprehension of factors that shape entrepreneur's perception in selecting the adequate instrument of finance for their businesses. Previewing Scholars argued that individual and contextual factors influence entrepreneurs' financing decision-making (Atherton 2012; Nouira et al. 2005). Our research concerns the manner in which these factors are important for choice orientation despite financing options. In our analysis, we classified the factors into two blocks, impeding and enabling factors.

Impeding factors constrain and limit the availability of financing options, which arise from both contextual and individual levels. In contrast to impeding factors, enabling factors allow entrepreneurs to identify and understand the financing options available in the market and thus lead them to decide the appropriate choice of financing for their start-up. According to our analysis, enabling factors are mainly individual aspects embodied in the entrepreneur's cognitive system. Surprisingly, we did not recognize any positive perception regarding the contextual factors by the respondents.

Our results show that lack of formation and risk aversion are the main individual factors that restrict the financing choice that entrepreneurs face. The literature analysis confirms the fact that specific human capital such as financial education and management knowledge are accepted as the key elements that support an adequate decision-making by entrepreneurs (Hussain et al. 2018; Seghers et al. 2012). We believe that nascent entrepreneurs with no financial literacy might experience a certain aversion toward risk as they cannot interpret correctly or act on them in an effective manner (Adomako et al. 2016).

Second, given the recent ecosystem's dynamics and the increasingly complex financial system, entrepreneurs are considered to be financially underinformed. Consequently, they develop negative emotions (anger, disappointment, and dissatisfaction) against the context in which they interact. Entrepreneurs, then, fail to grasp the complexity of sophisticated financial alternatives available for start-ups. As a result, start-up founders may favor simple and familiar sources of financing rather than complex ones (Matthews et al. 2013).

Third, in accordance with previous studies on entrepreneurship, our results demonstrate that cultural aspects could also impact the financial choices of an entrepreneur. Religious principles are intimately linked to the mental decision system of the entrepreneur in Islamic countries. In this sense, access to financing capital is strongly influenced by religious principles (Almobaireek et al. 2017).

In parallel with the impeding view, which limits and constrains the availability of external financing choices, the entrepreneur can compensate and benefit from some individual factors that may orient their choices. In line with previous studies, our research shows that the network is a key individual characteristic that helps the entrepreneur to evaluate and access financial resources (Heuven et Groen 2012; Rasmussen et Sørheim 2012). Hence, networks can decrease the ambiguity of different financial alternatives and reduce the asymmetry of information (Brown et al. 2018).

In addition, entrepreneurs can accommodate risk and take proactive action to alleviate it. Those entrepreneurs are more predisposed to choose the appropriate financial resource that fit the most with their strategic goals. Research on risk-taking is not consistent, some researchers consider that entrepreneurs are in nature less risk-averse than others (Scherr, Sugrue, et Ward 1993). Some studies confirm empirically that young entrepreneurs are more reluctant in taking risks than nonentrepreneurs (Xu et Ruef 2004).

Another interesting finding is the incentive role of self-achievement; we think that this aspect will have significant implications for how start-ups' owners respond to potential external investors. This result corroborates Sapienza's (2003) affirmations that the incentive for self-determination may drive the start-ups' owners to look for adequate entrepreneurial financing choices to achieve their nonpecuniary goals (Sapienza, Korsgaard, et Forbes 2003).

REFERENCES

Adler, P S. et S W Kwon. 2002. Social capital: Prospects for a new concept. *Academy of Management Review* 27(1):17–40.

Adomako, S, A Danso, et J O Damoah. 2016. The moderating influence of financial literacy on the relationship between access to finance and firm growth in Ghana. *Venture Capital* 18(1):43–61. Consulté http://www.tandfonline.com/doi/full/10.1080/13691066.2015.1079952.

Ahlstrom, D et G D Bruton. 2002. An institutional perspective on the role of culture in shaping strategic actions by technology-focused entrepreneurial firms in China. *Entrepreneurship Theory and Practice* 26(4):53–68. Consulté http://journals.sagepub.com/doi/10.1177/104225870202600404.

Almobaireek, W N, A A. Alshumaimeri, et T S Manolova. 2017. The process of new venture creation in the islamic world: an organizing framework. In *Entrepreneurship and Management in an Islamic Context*. Cham: Springer International Publishing. pp. 49–74. Consulté http://link.springer.com/10.1007/978-3-319-39679-8_5.

Atherton, A. 2012. Cases of start-up financing. *International Journal of Entrepreneurial Behavior & Research* 18(1):28–47. Consulté http://www.emeraldinsight.com/doi/10.1108/13552551211201367.

Bellavitis, C, I Filatotchev, D S Kamuriwo, et T Vanacker. 2017. Entrepreneurial finance: New frontiers of research and practice. *Venture Capital* 19(1–2):1–16. Consulté https://www.tandfonline.com/doi/full/10.1080/13691066.2016.1259733.

Block, J H, M G Colombo, D J Cumming, et S Vismara. 2018. New players in entrepreneurial finance and why they are there. *Small Business Economics* 50(2):239–50. Consulté http://link.springer.com/10.1007/s11187-016-9826-6.

Braunerhjelm, P. 2013. Entrepreneurship, innovation and economic growth: interdependencies, irregularities and regularities. In Handbook of Research on Innovation and Entrepreneurship. Edward Elgar Publishing. Consulté http://www.elgaronline.com/view/9781848440876.00021.xml.

Brown, R, S Mawson, et A Rowe. 2018. Start-ups, entrepreneurial networks and equity crowdfunding: A processual perspective. *Industrial Marketing Management* 80:115–25. Consulté https://linkinghub.elsevier.com/retrieve/pii/S0019850118300804.

Brush, C G, L F Edelman, et T S Manolova. 2012. Ready for funding? Entrepreneurial ventures and the pursuit of angel financing. *Venture Capital* 14(2–3):111–29. Consulté http://www.tandfonline.com/doi/abs/10.1080/13691066.2012.654604.

Bruton, G D, D Ahlstrom, et K Singh. 2002. The impact of the institutional environment on the venture capital industry in Singapore. *Venture Capital* 4(3):197–218.

Bruton, G D, V H Fried, et S Manigart. 2005. Institutional influences on the worldwide expansion of venture capital. *Entrepreneurship: Theory and Practice* 29(6):737–60.

De Carolis, D M et P Saparito. 2006. Social capital, cognition, and entrepreneurial opportunities: A theoretical framework. *Entrepreneurship Theory and Practice* 30(1):41–56. Consulté http://journals.sagepub.com/doi/10.1111/j.1540-6520.2006.00109.x.

Cassar, G 2004. The financing of business start-ups. *Journal of Business Venturing* 19(2):261–83. Consulté http://linkinghub.elsevier.com/retrieve/pii/S0883902603000296.

Chin, J T 2020. Location choice of new business establishments: Understanding the local context and neighborhood conditions in the United States. *Sustainability* 12(2):501. Consulté https://www.mdpi.com/2071-1050/12/2/501.

Coleman, S, C Cotei, et J Farhat. 2016. The debt-equity financing decisions of U.S. startup firms. *Journal of Economics and Finance* 40(1):105–26. Consulté http://link.springer.com/10.1007/s12197-014-9293-3.

Corbin, J et A Strauss. 2007. *Basics of Qualitative Research: Techniques and Procedures for Developing Grounded Theory*. 3rd éd. SAGE Publications, Inc.

Denzin, N K et Y S Lincoln. 2018. *SAGE Handbook of Qualitative Research*. 5th Ed. SAGE Publications, Inc.

Eisenhardt, K. M. 1989. Building theories from case study research. *Academy of Management Review* 14(4):532–50.

Eisenhardt, K M et M E Graebner. 2007. Theory building from cases: Opportunities and challenges. *Academy of Management Journal* 50(1):25–32.

Ekanem, I, R Owen, et A Cardoso. 2019. The influence of institutional environment on venture capital development in emerging economies: The example of Nigeria. *Strategic Change 28*(1):95–107.

Frid, C J, D M Wyman, W B Gartner, et D H Hechavarria. 2016. Low-wealth entrepreneurs and access to external financing. *International Journal of Entrepreneurial Behavior & Research* 22(4):531–55. Consulté 9 février 2019 http://www.emeraldinsight.com/doi/10.1108/IJEBR-08-2015-0173.

Ghauri, P N et R Firth. 2009. The formalization of case study research in international business. *der markt* 48(1–2):29–40. Consulté http://link.springer.com/10.1007/s12642-009-0003-1.

Gioia, D A, K G Corley, et A L Hamilton. 2013. Seeking qualitative rigor in inductive research. *Organizational Research Methods* 16(1):15–31.

Gompers, P et J Lerner. 2010. Equity financing. In *Handbook of Entrepreneurship Research*, édité par D B Audretsch et Z J Acs. New York, NY: Springer. Consulté http://link.springer.com/10.1007/978-1-4419-1191-9_8.

Granovetter, M. 1985. Economic action and social structure: The problem of embeddedness economic action and social structure: The problem of embeddedness. *American Journal of Sociology* 91(3):481–510.

Hall, B H 2002. The financing of research and development. *Oxford Review of Economic Policy* 18(1):35–51. Consulté https://academic.oup.com/oxrep/article-lookup/doi/10.1093/oxrep/18.1.35.

Hall, B H et J Lerner. 2010. The financing of R&D and innovation. *Handbook of the Economics of Innovation* 1(1C):609–39.

Heuven, J et A Groen. 2012. The role of social networks in financing technology-based ventures: An empirical exploration. *Venture Capital* 14(2–3):131–49. Consulté http://www.tandfonline.com/doi/abs/10.1080/13691066.2012.659473.

Hogan, T et E Hutson. 2005. What factors determine the use of venture capital? evidence from the Irish software sector. *Venture Capital* 7(3):259–83. Consulté http://www.tandfonline.com/doi/abs/10.1080/13691060500268249.

Hsu, D H 2007. Experienced entrepreneurial founders, organizational capital, and venture capital funding. *Research Policy* 36(5):722–41. Consulté https://linkinghub.elsevier.com/retrieve/pii/S0048733307000716.

Hussain, J, S Salia, et A Karim. 2018. Is knowledge that powerful? Financial literacy and access to finance. *Journal of Small Business and Enterprise Development* 25(6):985–1003. Consulté https://www.emeraldinsight.com/doi/10.1108/JSBED-01-2018-0021.

Julien, P-A. 1990. Vers une typologie multicritère des PME. *Revue internationale P.M.E.: Économie et gestion de la petite et moyenne entreprise* 3(3–4):411. Consulté http://id.erudit.org/iderudit/1007988ar.

Langley, A. 1999. Strategies for theorizing from process data. *Academy of Management Review* 24(4):691–710.

Luger, M I. et J Koo. 2005. Defining and tracking business start-ups. *Small Business Economics* 24(1):17–28. Consulté http://link.springer.com/10.1007/s11187-005-8598-1.

Mac an Bhaird, C et B Lucey. 2014. Culture's influences: An investigation of inter-country differences in capital structure. *Borsa Istanbul Review* 14(1):1–9. Consulté https://linkinghub.elsevier.com/retrieve/pii/S2214845013000094.

Marchesnay, M. 1991. La PME: une gestion spécifique. *Économie Rurale* 206(1):11–17.

Matthews, C H, M T Schenkel, M W Ford, et S E Human. 2013. Financing complexity and sophistication in nascent ventures. *Journal of Small Business Strategy* 23(1):15–30.

McKeever, E, A Anderson, et S Jack. 2014. Entrepreneurship and mutuality: Social capital in processes and practices. *Entrepreneurship and Regional Development* 26(5–6):453–477.

Megginson, W L 2004. Toward a global model of venture capital? *Journal of Applied Corporate Finance* 16(1):89–107. Consulté http://doi.wiley.com/10.1111/j.1745-6622.2004.tb00599.x.

Miles, M B, M Huberman, et J Saldaña. 2014. *Qualitative Data Analysis: A Methods Sourcebook*. 4th éd. SAGE Publications Inc.

Nouira, S, M Klofsten, et Å L Dahlstrand. 2005. The logic of the entrepreneur. *The International Journal of Entrepreneurship and Innovation* 6(2):85–95. Consulté 12 janvier 2019 http://journals.sagepub.com/doi/10.5367/0000000053966867.

Oviatt, B M. et P P McDougall. 1995. Global start-ups: Entrepreneurs on a worldwide stage. *Academy of Management Perspectives* 9(2):30–43. Consulté http://journals.aom.org/doi/10.5465/ame.1995.9506273269.

Pellegrino, G, M Piva, et M Vivarelli. 2011. How do young innovative companies innovate? In *Handbook of Research on Innovation and Entrepreneurship*, Édité par D B Audretsch, O Falck, S Heblich, et A Lederer. Edward Elgar. Consulté http://irx.sagepub.com/content/26/2/208%5Cnhttp://irx.sagepub.com/content/26/2/208.short.

Picken, J C 2017. From startup to scalable enterprise: Laying the foundation. *Business Horizons* 60(5):587–95. Consulté https://linkinghub.elsevier.com/retrieve/pii/S0007681317300605.

Rasmussen, E et R Sørheim. 2012. Obtaining early-stage financing for technology entrepreneurship: Reassessing the demand-side perspective. *Venture Capital* 14(2–3):77–89. Consulté 11 août 2018 http://www.tandfonline.com/doi/abs/10.1080/13691066.2012.667908.

Ren, Q, Y Xu, R Zhou, et J Liu. 2020. Can CEO's humble leadership behavior really improve enterprise performance and sustainability? A case study of Chinese start-up companies. *Sustainability* 12(8):3168. Consulté https://www.mdpi.com/2071-1050/12/8/3168.

Sapienza, H J, M A Korsgaard, et D P Forbes. 2003. The self-determination motive and entrepreneurs' choice of financing In *Cognitive Approaches to Entrepreneurship Research*, vol. 6, édité par J A Katz et D A Shepherd. Elsevier Science Ltd. Consulté https://www.emeraldinsight.com/10.1016/S1074-7540(03)06005-7.

Scherr, F C, T F Sugrue, et J B Ward. 1993. Financing the small firm start-up: Determinants of debt use. *Journal of Small Business Finance* 3(1):17–36.

Seghers, A, S Manigart, et T Vanacker. 2012. The impact of human and social capital on entrepreneurs' knowledge of finance alternatives. *Journal of Small Business Management* 50(1):63–86. Consulté http://doi.wiley.com/10.1111/j.1540-627X.2011.00344.x.

Shepherd, D A, T A Williams, et H Patzelt. 2015. Thinking about entrepreneurial decision making. *Journal of Management* 41(1):11–46. Consulté http://journals.sagepub.com/doi/10.1177/.

St-Pierre, J et N Fadil. 2016. Finance entrepreneuriale et réalité des PME : une enquête internationale sur les connaissances et les pratiques académiques des chercheurs. *Management International* 20(2):52. Consulté 20 février 2019 http://id.erudit.org/iderudit/1046562ar.

Suddaby, R, G D Bruton, et S X Si. 2015. Entrepreneurship through a qualitative lens: Insights on the construction and/or discovery of entrepreneurial opportunity. *Journal of Business Venturing* 30(1):1–10.

Van Auken, H E. 2005. A model of small firm capital acquisition decisions. *The International Entrepreneurship and Management Journal* 1(3):335–52. Consulté http://link.springer.com/10.1007/s11365-005-2599-z.

Vanacker, T, A Heughebaert, et S Manigart. 2014. Institutional frameworks, venture capital and the financing of european new technology-based firms. *Corporate Governance: An International Review* 22(3):199–215. Consulté http://doi.wiley.com/10.1111/corg.12046.

Witte, P, B Slack, M Keesman, J H Jugie, et B Wiegmans. 2018. Facilitating start-ups in port-city innovation ecosystems: A case study of Montreal and Rotterdam. *Journal of Transport Geography* 71:224–234.

Xu, H et M Ruef. 2004. The myth of the risk-tolerant entrepreneur. *Strategic Organization* 2(4):331–55. Consulté http://journals.sagepub.com/doi/10.1177/1476127004047617.

Yin, R K 2009. *Case Study Research Design and Methods*, 4th ed, SAGE Publications, Inc.

Zahra, S A, M Wright, et S G Abdelgawad. 2014. Contextualization and the advancement of entrepreneurship research. *International Small Business Journal* 32(5):479–500.

4 Green Entrepreneurship
An Avenue for Innovative and Sustainable Product Development and Performance

Chandan Veer and Pavnesh Kumar
Mahatma Gandhi Central University, India

Ravindra Sharma
Swami Rama Himalayan University, India

CONTENTS

4.1 INTRODUCTION

The contribution of entrepreneurs to the development of society is significant and well-acknowledged (Schumpeter, 1934; Dean and McMullen, 2007). It is quite conflicting that the same business activities have been extensively blamed for the negative outcome or impact on the environment because of several business activities. These unsustainable or environmentally irresponsible business activities are called by researchers as an "old business model" (Mrkajic et al., 2019: Demirel et al., 2019). Several organizations including United Nations stressed on training and awareness programs among individuals for them to understand sustainable business practices (Pacheco et al., 2010). These mediating efforts aim to expand the cognizance of sustainable entrepreneurial actions, which led to Stockholm Conference

DOI: 10.1201/9781003217084-4

and Rio-De Janeiro Earth Summit, Kyoto Protocol, and several other similar events. Entrepreneurs will have to realize and move from traditional business activities, which are mainly based on profit at the cost of environment deterioration, and this is where the green entrepreneurship originates (Farinelli et al., 2013). The Organization for Economic Cooperation and Development recommends the policy framework through green entrepreneurship to address global environmental friendliness (OECD, 2011). The unsustainable traditional business practices have obstructed the aspiration to attain globally a lesser carbon-generating economy, hence increasing the threat of the earth's prosperity (O'Neill and Gibbs, 2016). This shows the incapability of business persons to practice their business activities under the umbrella of green or sustainable entrepreneurship, failing to encash the opportunities of green entrepreneurship. Green entrepreneurs can have an advantage over old business practices by identifying and fulfilling the gap of the emerging green markets (Dean and McMullen, 2007). As a green entrepreneur, it is equally important to identify the prospects and challenges associated with green entrepreneurship (Esty and Winston, 2009). This has begun a new era of business having a green outlook. Green ventures focus on green innovation, efficiency, market catering, higher revenue and growth, calculated risk, and relationship marketing (Porter and van der Linde, 1995). The unsustainable business practices contribute to environmental pollution, destructing the natural environment, and entrepreneurs are expected to be ready to face the consequences of their business practices in this regard. Sustainable development has always attracted attention and has been treated as an important issue in any session, be it enterprise, regulatory body, or society (Kulkarni and Pammar, 2019; Rana et al., 2017). Sustainability is elucidated as a development, which satisfies the present-generation needs without compromising the capacity to fulfill the future generation needs (WCED Commission, 1987). Similarly, sustainable development is a present-time approach to development that maintains a balance among environment, society, and economy (Rahman et al., 2016). Similarly, Hussain (2018) expressed that the sustainability of environment is gaining attention. But entrepreneurs engaged in green business and capable of contributing toward socio-environmental development is still ambiguous.

The increased attention toward sustainable development through green entrepreneurship led to the innovation and development of green products (Rana and Sharma, 2021). In the past few decades, organizations and enterprises have initiated to inculcate ecological aspects in their product innovation and development strategies, which has resulted in the growth of a number of green products. Such a paradigm shift is due to the growing awareness toward green products, which signifies an attractive market potential (Seebode et al., 2012). Organizations that strive to launch a green product face several challenges (Dangelico and Pujari, 2010); therefore, it is necessary to consider the interest of the organization and have a proper understanding while developing green products. A better insight into ecological new product development (NPD) is the focal point for the improvement in the process and minimizing the nonsuccess of green product innovation (Dangelico and Pujari, 2010). Currently, we notice the accelerating change in consumers' taste and preferences, and the shrinking of product life cycle. This challenge is crucial for corporates to effectively frame NPD (Calantone et al., 1995). The nucleus of effective NPD depends upon its core attributes, that is, its basic benefits and auxiliary attributes, that is, the features that differentiate it from other products. There are several attributes as per the expectation of stakeholders, customers, consumers, or government groups. Choosing any suitable set of features is attained by reducing the process called "Design for F", where "F" signifies a proper set of features such as quality, safety, and ease of use (Gatenby and Foo, 1990). An important factor that has originated in the last 10 years is the "sketch for environment," which is defined as "a practice by which environment credentials are incorporated into product, process, and design procedures" (Keoleian et al., 1995). In the car industry, manufacturers emphasize on environmental consideration during the product development process with a pursuit for energy-efficiency and low carbon emission (Thornton, 1999). In the United States, a percentage rise in green products from 1.1% to 13.4% was observed among the new products that were introduced (Ottman, 1998). Though the strict regulation and consumer misbelief have reduced the introduction of such new products, the market for products such as paper towel, dishwasher liquid, and other products is growing steadily as a green brand (Speer, 1997). The overall green products in the US market accounted for 10%, with 30% of the green products in household category (Fuller, 1999). This paradigm shift is not only limited to the manufacturing industry but is also extended to the tourism and financial services market. These issues arose because the environmental consideration has led to several changes in the aspects of business operation, purchasing behavior, marketing, and organizational strategy (Wong et al., 1996; Shrivastava and Hart, 1994; Rana and Sharma, 2019; Menon and Menon, 1997). Earlier, the environment-related response that was treated as mere adherence, extra cost, or tradeoff in the corporate goals is now viewed as a good chance. Like a winning strategy for being both sustainable and having a competitive advantage (Porter and van der Linde, 1995; Rana et al., 2019) with some different opinions (Walley and Whitehead, 1994), the extensive literature depicts other advantages apart from environmental success such as increased revenue (Fierman, 1991), improved customer satisfaction (Frankel, 1992), companionship with customers (Dean et al., 1995), increased competition (Miles and Munilla, 1993; Porter and van der Linde, 1995), and strong corporate image (Engleberg, 1992; Kolk, 2000). Menon and Menon (1997) and many business leaders believe that environmental consideration and business success are both complementary and are associated together termed as "Ecopreneuring." There is a link between environment, productivity, innovation, and competitiveness. The marketing strategies should be framed focusing on both environmental aspects and organizations' economic aspects (Porter and van der Linde, 1995; Ottman, 1994). Product caring factors comprise all aspects from the products' performance to the products' life cycle. The development of products that is environmentally better is crucial for both the company and the environment. In spite of knowing the fact that it is important to integrate the environmental aspects in the NPD process, the literature available is informal and based on unscientific proof. In this chapter, we understand the aspects of ecological (environmental) new product development (ENPD) by reviewing the literature and suggesting a conceptual framework of ENPD. The ecopreneurship has a root in ENPD. The effectiveness and success of any green entrepreneurship depends upon how sustainable its products or services are in terms of market offerings.

4.2 REVIEW OF LITERATURE

4.2.1 Concept of Green Entrepreneurship

In spite of expeditious development in the field of entrepreneurship, green entrepreneurship, as an area of research, is still in its initial stage. Pacheco et al. (2010) express that some related fields such as entrepreneurship, business-related economics, financial aspects, and accounting measures should to be researched. Entrepreneurs still face dilemma and hypothesize identifying factors that represent green entrepreneurship (O'Neill and Gibbs, 2016). To eradicate this state of confusion, researchers from time to time at several occasions have explained the concept of green entrepreneurship for better understanding. Green entrepreneurship is an activity that deliberately addresses environmental and social issues through executing green entrepreneurial strategies with an expectation of constructive environmental response and financial gain (Green Project, 2012). Green entrepreneurs are those who run their ventures designed to be eco-friendly in process and products. Green entrepreneurship is an adopted line of technology-based production or organization activities (Sunny and Shu, 2017). Green entrepreneurship is a storytelling process by convincing the stakeholders and thus the entrepreneur realizes his/her objectives (Dale, 2018). The lack of common consent among researchers for green entrepreneurship is awaiting a globally accepted explanation (Demirel et al., 2017). In the business industry, 60% of businesses are calculating their effectiveness and efficiency through sustainable programs from which 78% are power efficient, two-thirds save paper, and 60% have reduced excessive water consumption (Buck Consultants, 2011). Similarly, Khan (2015) argues that the unique answer to sustainable challenges for entrepreneurs is to align their business activities and depend on the "Green Movement" for long-term survival and prosperity. Certain business houses practice "Greenwashing" to attract customers for their brand by marketing tricks, but now they have realized the strength of green practices for spreading their consumer base for their brand. The accurate green enterprises in spite of growing hindrances through greenwashing are enjoying a high degree of loyal consumer base (Darnall and Edwards, 2006; Zhaojun et al., 2017). The unethical business houses, which claim to be proenvironmental or eco-friendly for gaining undue market advantage, can only have a short-term benefit. Business organizations that have realized the importance of green strategies are gearing up to position themselves for sustainable market opportunities. Those who pretend to be green are posing a greater challenge to young green entrepreneurs, which cannot be ignored. These challenges arise due to inadequate government regulations. Green entrepreneurship is an area that can protect environment deterioration and is completely related to sustainable business activities (Ndubisi and Nair, 2009). Crucial areas of business are being negatively influenced by nature-related issues. The incorporation of green strategies that improve the reputation of the organization and provide benefits to investors should be made the core of the organizational business practices (Coddington, 1993; Banerjee, Charles, and Easwar, 1995). To mitigate these environmental challenges within the scope of "Green" is important for "continuation and profitability" for every business house. (Baker and Sinkula, 2005). The green attitude and perception toward business give long-lasting support for the establishment and prosperity of a business (Schaper, 2010). The corporate social responsibility (CSR) activities are directly influenced by organizations' internal activities and environment responsiveness (Shah, Arjoon, and Rambocas, 2016). Individual sustainable practices have a substantial contribution through green technology and development (Chen, 2019). The factors related to philanthropy, time viewpoint, and association were identified as the most impacting factors for ecopreneurs (Vatansever and Arun, 2016). In a similar manner, Braun's (2010) study showed that women are more environmentally committed with a positive attitude toward the green entrepreneurship program compared to men. The shortage of funds, limited access to education, and lack of innovativeness are found to be the major hurdles that female entrepreneurs face in commencing any green venture (Bergset, 2018). A country's internal factors and government rules considerably impact the entrepreneurship (Aslam and Hasnu's, 2016). According to Giddings et al. (2002), when there is a separation in the environment, society, and economy, it results in limited technology-based development. There is a strong association between proenvironmental activities and the performance of the organization (Baker and Sinkula, 2005). The term "Green" is applied to the actions initiated to minimize the environmental deterioration and to carry out environmental sustainability (Parry, 2012). Entrepreneurial thoughts play an important role in constructing a green business model having resemblance with the government policies (Hockerts and Wustenhagen, 2010). There is a challenge to balance among the triple bottom line, that is, profit, the planet, and people. To stimulate sustainable enterprises, motivation act as an important factor (Masurel, 2007). These factors have a direct role in fostering ecopreneurship, which enhances the green economy. Ecopreneurship has an important role in fostering individuals to embrace ideas of sustainable businesses (Pandrani and Ferguson, 2013). The youth appears more passionate about taking entrepreneurship by initiating their own business (Shah and Soomro, 2017; Sharma and Rana, 2021). Sustainability is defined as the protection of productivity to meet the endless demand, the maintenance of desired life, and conserving the environment (Li, Okoroafo, and Gammoh, 2014; Prakash et al., 2021). Educational programs have a strong correlation with the entrepreneurial attitudes of newly evolved entrepreneurs (Schroder and Schmitt-Rodermund, 2006). Green entrepreneurship has a positive and substantial relation with sustainability orientation and education (Soomro et al., 2019). The educational program framed properly can develop entrepreneurial intentions and attitudes among the potential entrepreneurs through inspiration, motivation, and learning (Souitaris, Zerbinati, and Al Laham, 2007). It is presupposed that sustainability education can trigger the reorientation in the individuals' mind for entrepreneurship (Sterling, 2001). Several researchers propounded that sustainable education would motivate the citizens to transform to learn and have a sustainable world (Wals, 2009).

4.2.2 Environmental New Product Development

It is observed that NPD had earlier directed attention toward environmental challenges, which further shifted attention toward green NPD in many environment-related literature reports (Albino et al., 2009; Baumann et al., 2002; Handfield et al., 2001).

There is lack of sufficient information and knowledge as to why and how organizations should integrate environmental sustainability in the process of NPD (Dangelico and Pujari, 2010). Whether any product has green attributes is a complex issue and is a matter of debate (Dangelico and Pujari, 2010; Schiederig et al., 2012). Products that declare to have 'environmental advantages' are generally not so, rather they are less harmful to the nature compared to others (Ottman et al., 2006). The green products are defined as a new products having attributes that are considerably better than traditional or competitive products. Green attributes of a product are a continuous process rather than dichotomous (Schiederig, Tietze, and Herstatt, 2012). Green products reflect a considerable enhancement in product features, which could be small or big in terms of environment benefits, whereas nongreen or traditional products refer to no or negligible upgradation in products. Green products have several hidden attributes (Mangun and Thurston, 2002; Dangelico and Pontrandolfo, 2010). The impact of any product on environment signifies three dimensions: nature of materials, energy usage, and pollution generated (Dangelico and Pontrandolfo, 2010). In all of the above-stated dimensions, the impact on environment can be minimized and a significant contribution toward the environment can be observed (Dangelico and Pontrandolfo, 2010). The companies who want to utilize the opportunity should develop green products, and for this, they must embrace proactive, strategic actions to mitigate green challenges like climate change (Porter and Reinhardt, 2007). It is also observed that there is a rise in the number of firms embracing European Eco-labels for their products (European Commission, 2010). European Commission defines green products as "products (that) consumes lesser inputs or resources, having lesser impact and risk toward environment and reduce waste generation already at initial stage." Likewise, products that provide high levels of environmental benefits or having lesser environmental cost in comparison to similar products are green products (Reinhardt, 1998). Green products involve innovation in technologies, which comprises reduction in energy consumption, lesser pollution, recycling of waste, eco-friendly design, and corporate environmental management (Chen et al., 2006). The academic research suggests descriptive instructions for engineer and managers to incorporate environmental factors during NPD, while academic studies have failed to attract attention on existing theoretical framework of NPD (Berchicci, 2009). When environmental impacts are considered in NPD, they are treated as unimportant because of the cost and time involved (Handfield et al., 2001). There is a substantial gap between the incorporation of environmental affairs in any product or service and the well-accepted NPD literature (Baumann et al., 2002).

4.3 OBJECTIVES OF THE STUDY

- To propose the ENPD process model with a view to simplify few unresolved concepts.
- To explore various perspectives of green entrepreneurship through an integrative review of the literature.
- To analyze the impact of ENPD on green entrepreneurship.

4.4 RESEARCH METHODOLOGY

The methodology adopted for the current study was an integrative review of literature. Many elements such as environmental product policy, eco-friendly product material, product testing, and suppliers' involvement were identified. Researchers tried to explore the past and conceptual models of green NPD and ecopreneurship through a systematic search of the existing Internet-based manuscripts and hard copy publication of the respective fields.

4.5 ENVIRONMENTAL NPD PROCESS MODEL

The traditional NPD decision-making process (Kotler, 2003, p. 335) mainly focuses on the perspective of commercialization and profit. Its area of operation includes market, consumer, and profitability. The traditional NPD process begins with the idea generation involving innovation and creativity as important factors. The further processes include idea screening, concept development and testing, marketing strategy development, business analysis, product development, market testing, and finally commercialization, subject to adhering viability and profitability (Kotler, 2003, p. 335). The current environmental deterioration has stressed to incorporate sustainable practices or eco-friendly strategies while developing a new product. The final product thus manufactured, considering the environmental factors, will be termed as green products or environmentally friendly products. Remarkably, few studies have identified the impact of the performance of green products (Pujari and Wright, 1996; Prothero and McDonagh, 1992; Ryding, 1994; Chao-Tung, 1994). There is a lack of systematic studies that can explain the process that the business undergoes during manufacturing while adhering to environmental concerns (Buchholz et al., 1995). The environment-oriented product, in order to succeed, requires effective marketplace performance. Incorporating the latest environmental technology will not help the pursuance of sustainability unless they communicate it properly and gain the market share from traditional products. Environmental NPD is a process intended to transform the intention, attitude, and future actions of a business interested in the process of incorporating innovation in products (Crul, 1994). Several concepts of the NPD process have been suggested, examined, and incorporated but the environmental part has always been neglected (Urban and Hauser, 1980; Wind, 1982). Consumers' dilemmas are increasing day by day about all the products that are marketed on environmental factors but in reality have low environmental attributes (Gray-Lee et al., 1994).

Figure 4.1 summarizes the key factors to be considered or to be taken into account for the ENPD process.

The environmental product development starts with the raw material procurement in which the entrepreneur should evaluate the nature of the procured raw material. The locally procured raw material always has a low impact on the environment. It is always not practically applicable to procure materials locally or from nearby area as they may be subjected to availability. But the firm should strive to scrutinize the possible options available for material procurement. The composition of the raw material should be organic and eco-friendly. The organic or eco-friendly raw materials

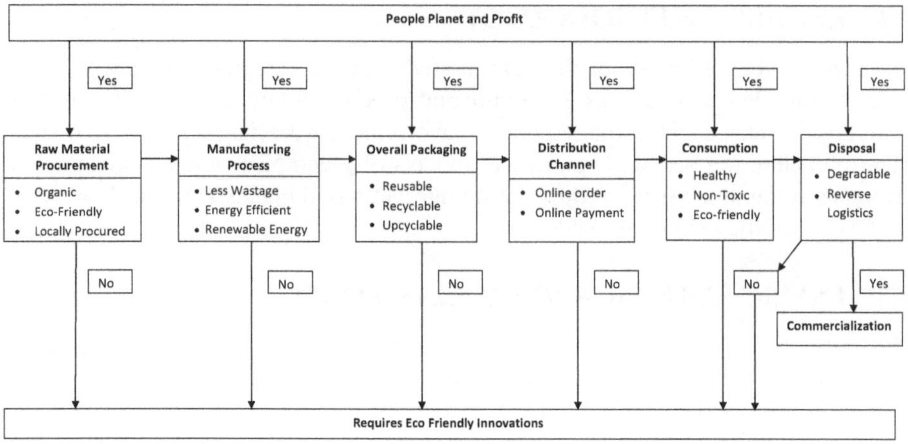

Source: Author's own

FIGURE 4.1 Environmental new product development process.

signify that they will have a minimal negative impact on the health of humans and the environment. The case of organic or eco-friendly raw material can be chemically free, nontoxic, nongenetically modified, and so on.

The second step of ENPD is the manufacturing process adopted by the business firm. The manufacturing process should ensure zero defects and low wastage. Zero defects and zero effect should be the objective of the manufacturing process. The environmentally positive and innovative technology of manufacturing should be embraced by the entrepreneurs for manufacturing of products. The use of renewable energy should always be the prime focus for the functioning of the plants and machineries. The entrepreneur should strive to be energy efficient, as small amounts of energy-saving will substantially contribute to the betterment of the environment. A similar approach should be used toward the consumption of water for manufacturing process.

The third stage is all about packing and packaging of the final manufactured goods. The materials nowadays used for packing and packaging are environment-friendly. The packing and further packaging of the product is crucial and compulsory for the products' safety, keeping it away from contamination, avoid spilling, and to give a unique recognition through its brand name and descriptions. The entrepreneur should aim to use easily degradable material for the packing and packaging purposes. The entrepreneur should ensure whether the packing material is completely degradable or not, and if the packing material is not completely degradable, then it should be recyclable, reusable, or even better can be upcycled. The upcycle of packing materials means transforming it into other usage rather than dumping it or considering it for recycling process. The enterprise should consider packing and packaging as an important factor of ENPD process because the outer cover of the products, in the initial stage, is visualized and observed by the consumers during their purchase decisions. The first impression gives a strong and unique recognition of the company, brand, and product itself.

The fourth step elucidates the distribution channel. In many cases, it has been identified that the distribution channel is something, which is not owned or fully controlled by the organization itself. They collaborate or hire the stakeholders for the distribution of the goods. These distributors should also be like-minded. In addition, they should ensure and attempt to be efficient in their operations such as online payment and order facility, a routine of distribution of goods.

The significance of ENPD does not end once it reaches the consumers hand similar to the traditional NPD process. The fourth step deals with the consumption and its consequences on human health and environmental impacts. The entrepreneur should ensure that the finished product should be free from all kinds of health hazards and should be nontoxic, healthy, and safe for consumption. The consumption or usage of such product should not bring any negative consequences on human and environmental health.

The last stage stresses on the elements of disposal. The remaining items after consumption should either be completely biodegradable or at least be recyclable. If it is of recyclable material, then the provision of reverse logistics will give an added advantage or a strong ecological image to the organization.

If in all the above steps, the people, the planet, and profitability or adhering the parameters or factors associated with the environment are considered, then the entrepreneur can proceed with further plans and frame an effective and targeted communication strategies for their customers. If the following steps are inadequate or lacking the environmental credentials, then it should be considered to innovate and develop the product as per the environmental expected compliance.

4.6 THEORETICAL AND MANAGERIAL IMPLICATIONS

This study has direct involvement for entrepreneurs and ecological NPD. Specifically, it fixes a set of factors to check the significance of the ENPD for the ecopreneurs and provides a significant evaluating technique, which can assist in measuring the market performance as well as features and further performance of ENPD. The main theoretical application of the current study is the development of the concept-based framework of ENPD, which incorporates the NPD literature with ecological marketing and sustainable management. This concept comprises ENPD factors, associated challenges for entrepreneurs to give utmost attention while developing eco-products, and further its performance in the market. This study is important because there is a lack of any factor-based theory in ENPD. The current study thus comes up with a deeper understanding and insight of the associated factors of ecological concern in decision for the product development and ENPD final outcome. The development of the conceptual framework involves incorporation of the literature review and identification of the product factors that impact its environmental credentials. This study imparts further assistance to the existing theory related to entrepreneurship and key activities of ENPD process. Various factors that are the keys to success in ENPD would also have concerns with the traditional product and traditional business practices. For the entrepreneurs who want to bring substantial improvement in their existing products as per the environmental norms, this study will benefit them effectively. To achieve better market results and efficient eco-performance from ENPD through improving inputs, reduce pollution and wastage, and increase environmental positivity, this study will contribute toward such aims and objectives. ENPD performance also depends on the cross-functional coordination and benchmarking. The validity of the NPD is rare, but ENPD requires a win–win situation for the ecopreneuring paradigm. The success warranty of the ENPD includes top management involvement and support. Formulating the environmental responsiveness in and organization through policies, regulations, and processes will positively influence the ENPD implementation and performance. This ENPD process becomes hampered if the organization has a conservative mindset and lack of technology.

This study elucidates the importance of ENPD and the success of ecopreneurs in business. The framework of NPD is still applicable in the ENPD, that is, achieving market success and environmental excellence simultaneously.

4.7 DISCUSSION AND CONCLUSIONS

At the global level, major attention is transferred to the ecopreneurship model or green business practices, owing to the fact that they have the ability to provoke sustainability. This paradigm shift was initiated because of the deteriorating nature of old business practices that cause unsustainability. The ecopreneurship results in the reduction of paper usage and other natural resources, stock consumption, rise in innovative technology adaption, recycling, and efficiency in energy consumption, which can be realized or implemented through ENPD. Needless to say, green entrepreneurship is a field that is making an appearance and establishing basic principles of process, procedures, and strategies for entrepreneurship management. The proper training and development for complete workforce and involvement of the top to floor management can initiate green behavior, which will further lead to green business activities in the firm. In order to realize this, ecopreneurs should learn to instill the ENPD idea in every member within and outside the establishment. The advantage of ENPD is that it directly benefits the green entrepreneurship whether it is a well-established organization or a newborn green enterprise, and its effort toward reducing the energy cost or other nonrenewable resources enhances sustainable practices and acquires free publicizing among staff, consumers, stakeholders, and shareholders that the organization is sensible toward nature and its resources.

This chapter concludes that there are enormous advantages of green entrepreneurship. The benefits include efficiency of the firm, social goodwill, and cost reduction. The ENPD process reflects a way for green entrepreneurs to incorporate the environmental concerns with the traditional NPD process with a paradigm. The emphasis of the study is to bring improvement in the traditional product and technologies and at the same time create new sustainable products. The ENPD process still requires a fundamental re-evaluation. The eco-performance of the product does not only depend on the attributes of the product but also on the production system and complete logistics and supply chain management. The authors strongly believe and acknowledge that there is an urgent need for more research in the field of ENPD and green entrepreneurship and its area of management.

REFERENCES

Albino, V., Balice, A., and Dangelico, R. M. (2009). Environmental strategies and green product development: An overview on sustainability-driven companies. *Business Strategy and the Environment*, 18(2), 83–96.

Aslam, S. and Hasnu, S. A. F. (2016). Issues and constraints perceived by young entrepreneurs of Pakistan. *World Journal of Entrepreneurship, Management and Sustainable Development*, 12(1), 50–65.

Baker, W. E. and Sinkula, J. M. (2005). Environmental marketing strategy and firm performance: Effects on new product performance and market share. *Journal of the Academy of Marketing Science*, 33(4), 461–475.

Banerjee, S., Charles, S. G., and Easwar, I. (1995). Shades of green: A multidimensional analysis of environmental advertising. *Journal of Advertising*, 24(2), 21–32.

Baumann, H., Boons, F., and Bragd, A. (2002). Mapping the green product development field: Engineering, policy and business perspectives. *Journal of Cleaner Production*, 10(5), 409–425.

Berchicci, L. (2009). Innovating for sustainability: Green entrepreneurship in personal mobility. *Innovating for Sustainability: Green Entrepreneurship in Personal Mobility*, 1–232. 10.4324/9780203889565.

Bergset, L. (2018). Green start-up finance – Where do particular challenges lie?. *International Journal of Entrepreneurial Behavior & Research*, 24(2), 451–575.

Braun, P. (2010). Going green: Women entrepreneurs and the environment. *International Journal of Gender and Entrepreneurship*, 2(3), 245–259.

Buchholz, R. A., Marcus, A. A., Post, J. E., (eds.) *Managing Environmental Issues: A Case Book*. Upper Saddle River, NJ: Prentice-Hall, 1995.

Buck Consultants. (2011). American businesses are measuring the savings from their green workplace practices more than ever. Retrieved on 14/6/2021 via http://news.xerox.com/pr/xerox/photo.aspx?fid=137069.

Calantone, R. J., Vickery, S. K., & Dröge, C. (1995). Business performance and strategic new product development activities: an empirical investigation. *Journal of Product Innovation Management: An International Publication of the Product Development & Management Association*, 12(3), 214–223.

Chao-Tung (Jordan), W. *Integrating Environmental Management into Product Development: Promises and Limits at McDonald's, Procter and Gamble, and Warner-Lambert.* Unpublished doctoral dissertation, Troy, NY: Rensselaer Polytechnic Institute, 1994.

Chen, Yu-Shan & Lai, Shyh-Bao & Wen, Chao-Tung. (2006). The Influence of Green Innovation Performance on Corporate Advantage in Taiwan. *Journal of Business Ethics*, 67, 331–339. 10.1007/s10551-006-9025-5.

Chen, S. (2019). True sustainable development of green technology: The influencers and risked moderation of sustainable motivational behavior. *Sustainable Development*, 27(1), 69–83.

Chen, Y. S. (2008). The driver of green innovation and green image – green core competence. *Journal of Business Ethics*. 81, 531–543. https://doi.org/10.1007/s10551-007-9522-1.

Coddington, W. (1993). *Environmental Marketing*. New York: McGraw-Hill.

Crul, M. R. M. (1994). Milieugerichte produktontwikkeling in de praktijk. *Promise*. 1994;151–7.

Dale, G. (2018).The emergence of an ecological Karl Marx: 1818–2018. *The Legacy of the Political Economist Is Fiercely Contested*, 1.7. Retrieved on 13/07/2019 via https://climatecommunismandtheageofaffluencereview.org/.

Dangelico, R. M. and Pujari, D. (2010). Mainstreaming green product innovation: Why and how companies integrate environmental sustainability. *Journal of Business Ethics*, 95(3), 471–486.

Darnall, N. and Edwards, D. (2006). Predicting the cost of environmental management system adoption: The role of capabilities, resources and ownership structure. *Strategic Management Journal*, 27(4), 301–320.

Dean, T. J., Fowler, D. M., and Miller, A. Organizational adaptations for ecological sustainability: A resource-based examination of the competitive advantage hypothesis. Working Paper, Department of Management, University of Tennessee, Knoxville (TN), 1995.

Dean, T. J. and McMullen, J. S. (2007). Toward a theory of sustainable entrepreneurship: Reducing environmental degradation through entrepreneurial action. *Journal of Business Venturing*, 22(1), 50–76.

Demirel, P., Cher Li, Q, Rentocchini, F. and Tamvada, J. P. (2017). Born to be green: New insights into the economics and management of green entrepreneurship. *Journal of Business Economics*, 52(1), 759–771.

Development (DESD). (2005–2014). *United Nations Educational, Scientific and Cultural Organization (UNESCO)*. Workshops of UNESCO.

Driessen, P. H., B. Hillebrand, R. A. W. Kok, and T. M. M. Verhallen (2013). Green new product development: The pivotal role of product greenness. *IEEE Transactions on Engineering Management*, 60(2), 315–26. https://doi.org/10.1109/TEM.2013.2246792.

Engleberg, D. (1992). Is this the best community relations program in the country. *Environmental Manager*, 3, 6.

Esty, D. C. and Winston, A. S. (2009). Green to gold: How smart companies use environmental strategy to innovate, create value, and build competitive advantage. *Journal of Building and Engineering*, 2(5), 1–13.

Farinelli, F., Bottini, M., Akkoyunlu, S., and Aerni, P. (2013). Green entrepreneurship: The missing link towards a greener economy. *ATDF Journal*, 8(3), 42–48.

Fierman, J. (1991). The big muddle in green marketing. *Fortune*, 123:91–101.

Frankel, C. (1992). Blueprint for green marketing. *American Demographics*, 14, 34–8.

Fuller, D. *Sustainable Marketing: Managerial–Ecological Issues*. Thousand Oaks, CA: Sage, 1999.

Gatenby, D. A. and Foo, G. (1990). Design for X (DFX): Key to competitive, profitable products. *AT&T Technical Journal*, 2–13(May–June).

Giddings, B., Hopwood, B., and O'Brien, G. (2002). Environment, economy and society: Fitting them together into sustainable development. *Sustainable Development*, 10(4), 187–196.

Gray-Lee, J. W., Scammon, D. L., Mayer, R. N. (1994). Review of legal standards for environmental marketing claims. *Journal of Public Policy & Marketing*, 13, 155–9.

Green Project. (2012). An attempt to define green entrepreneurship. *NCCR Policy Brief*, 6(1), 1 3. http://greentproject.eu/wp-content/uploads/2016/01/Definition-green-entrepreneurship.pdf.

Handfield, R. B., Melnyk, S. A., Calantone, R. J. and Curkovic, S. (2001). Integrating environmental concerns into the design process: The gap between theory and practice. *IEEE Transactions on Engineering Management*, 48(2), 189–208.

Hockerts, K. and Wustenhagen, R. (2010). Greening Goliaths versus emerging Davids – Theorizing about the role of incumbents and new entrants in sustainable entrepreneurship. *Journal of Business Venturing*, 25(5), 481–492.

Hussain, A. (2018). Green human resource management (HRM) practices in organizations: A comprehensive literature surveyed. *Journal of Management Research and Analysis*, 5(2), 112–117.

Keoleian, G. A., Kock, J. E., Menerey, D. (1995). *Life Cycle Design Framework and Demonstration Projects*, Washington, DC: U.S. Environmental Protection Agency.

Khan, M. (2015). Green human resource management – A prerequisite for sustainable environment. *Progress in Science and Engineering Research Journal*, 18(3), 1–7.

Kolk, A. (2000). Green reporting. *Harvard Business Review*, 78(1), 15–6.

Kotler, P. (2003). *Marketing Management* (11th ed.). Pearson Education, p. 335.

Kulkarni, P. V. and Pammar, L. S. (2019). A comparative study of rating systems in green building. *Journal of Engineering and Technology*, 6(4), 1.3.

Li, S., Okoroafo, S., and Gammoh, B. (2014). The role of sustainability orientation in outsourcing: Antecedents, practices, and outcomes. *Journal of Management and Sustainability*, 4(3), 27–36.

Mangun, D. and Thurston, D. L. (2002). Incorporating component reuse, remanufacture, and recycle into product portfolio design. *IEEE Transactions on Engineering Management*. 49(4), 479–490.

Masurel, E. (2007). Why SMEs invest in environmental measures: Sustainability evidence from small and medium-sized printing firms. *Business Strategy and the Environment*, 16(3), 190–201.

Menon, A. and Menon, A. (1997). Enviropreneurial marketing strategy: The emergence of corporate environmentalism as market strategy. *Journal of Marketing*, 61, 51–67.

Miles, M. P. and Munilla, L. S. (1993). The eco-orientation: An emerging business philosophy. *Journal of Marketing Theory and Practice*, 1, 43–51.

Montoya-Weiss, M. M. and Calantone, R. J. (1994). Determinants of new product performance: A review and metaanalysis. *Journal of Product Innovation Management*, 11(5), 397–417.

Mrkajic, B., Murtinu, S. and Scalera, V. G. (2019). Is green the new gold? Venture capital and green entrepreneurship. *Small Business Economics*, 52(4), 929–950.

Muo, I. and Azeez, A. (2019). Green entrepreneurship: Literature review and agenda for future research. *International Journal of Entrepreneurial Knowledge*, 7(2), 17–29. https://doi.org/10.12345-0007.

Ndubisi, N. O. and Nair, S. R. (2009). Green entrepreneurship (GE) and green value added (GVA): A conceptual framework. *International Journal of Entrepreneurship*, 13, 21–34.

O'Neill, K. and Gibbs, D. (2016). Rethinking green entrepreneurship – fluid narratives of the green economy. *Journal of Environment and Planning*, 48(9), 1727–1749.

OECD. (2011). Measuring green entrepreneurship. *Entrepreneurship at a Glance*, 23–38. Retrieved on 20/06/2021 via: https://doi.org/10.1787/9789264097711-4-en.

Ottman, J. (1998). *Green marketing: Opportunity for Innovation*. 2nd ed. Lincolnwood, IL: NTC/Contemporary Books.

Ottman, J. A. (1994). *Green Marketing: Challenges and Opportunities*. Lincolnwood, IL: NTC Business Books.

Ottman, J. A., Stafford, E. R., and Hartman, C. L. (2006). Avoiding green marketing myopia: Ways to improve consumer appeal for environmentally preferable products. *Environment*, 48(5), 22–36.

Pacheco, D. F., Dean, T. J., and Payne, D. S. (2010). Escaping the green prison: Entrepreneurship and the creation of opportunities for sustainable development. *Journal of Business Venturing*, 25(5), 464–480.

Pandrani, M. D., and Ferguson, D. (2013). Enabling factors of green entrepreneurship: A case study of organic agriculture produce in the Sindh and Balochistan provinces of Pakistan. *Entrepreneurial Learning*, 3(1), 181–190. https://hrcak.srce.hr/file/192539.

Parry, S. (2012). Going green: The evolution of micro-business environmental practices, business ethics. *A European Review*, 21(2), 220–237.

Porter, M. E. and Reinhardt, F. L. (2007). A strategic approach to climate. *The Harvard Business Review*, 85(10), 22–26.

Porter, M. E., and van der Linde, C. (1995). Green and competitive: Ending the stalemate. *Harvard Business Review*, 120–33(September–October).

Prakash, C., Saini, R., and Sharma, R. (2021). Role of internet of things (IoT) in sustaining disruptive businesses. In R. Sharma, R. Saini, C. Prakash, V. Prashad, *Internet of Things and Businesses in a Disruptive Economy* (1st ed.). New York: Nova Science Publishers.

Prothero, A. and McDonagh, P. (1992). Producing environmentally acceptable cosmetics? The impact of environmentalism on the United Kingdom cosmetics and toiletries industry. *Journal of Marketing Management*, 8(2), 147–66.

Pujari, D. and Wright, G. (1996). Developing environmentally-conscious product strategy (ECPS): A qualitative study of selected companies in Britain and Germany. *Marketing Intelligence and Planning*, 14(1), 19–28.

Pujari, D., Wright, G., and Peattie, K. (2003). Green and competitive. *Journal of Business Research*, 56(8), 657–671. https://doi.org/10.1016/s0148-2963(01)00310-1.

Rahman, A., Nezakati, H., Ho, J. A. and Ong, S. T. (2016). Integrating operations, human resource, marketing, and accounting and finance perspectives in sustainable supply chain research. *Economic Growth, Environmental Protection & Social Progress*, 11–231.

Rana, G. and Sharma, R. (2019). Emerging human resource management practices in industry 4.0. *Strategic HR Review*, 18(4), 176–181.

Rana, G. and Sharma, R. (2021). Employer branding: Attracting and retaining employees for sustainable development in disruptive economy. *World Review of Science, Technology and Sustainable Development*, 17(4), 319. https://doi.org/10.1504/wrstsd.2021.117839.

Rana, G., Sharma, R., and Goel, A. K. (2019). Unraveling the power of talent analytics: implications for enhancing business performance. In: Rajagopal, B. R. (eds.) *Business Governance and Society*. Cham: Palgrave Macmillan. https://doi.org/10.1007/978-3-319-94613-9_3.

Rana, G., Sharma, R. and Rana, S. (2017). The use of management control systems in the pharmaceutical industry. *International Journal of Engineering Technology, Management and Applied Sciences*, 5(6), 12–23.

Reinhardt, F. L. (1998). Environmental product differentiation: Implications for corporate strategy. *California Management Review*, 40(4), 43–73. https://doi.org/10.2307/41165964.

Ryding, S. *International Experiences of Environmentally Sound Product Development Based on Life Cycle Assessment (LCA). Final report. AFR Report 36.* Stockholm: Swedish Waste Research Council, 1994.

Schaper, M. T. (2010). Understanding the green entrepreneur. In M. T. Schaper (ed.), *Making Ecopreneurs: Developing Sustainable Entrepreneurship*. Farnham: Gower Publishing, pp. 7–20. Retrieved from http://ezproxy.uws.edu.au/login?url=http://site.ebrary.com/lib/sydney/docDetail.action?docID=10395002.

Schiederig, T., Tietze, F., and Herstatt, C. (2012). Green innovation in technology and innovation management – an exploratory literature review. *R&D Manage*, 42(2), 180–192.

Schroder, E., and Schmitt-Rodermund, E. (2006). Crystallizing enterprising interests among adolescents through a career development program: The role of personality and family background. *Journal of Vocational Behavior*, 69(3), 494–509

Schumpeter, J. (1934). *The Theory of Economic Development*. Cambridge, MA: Harvard University Press.

Seebode, D., Jeanrenaud, S., and Bessant, J. (2012). Managing innovation for sustainability. *R&D Manage*. 42(3), 195–206.

Shah, K. U., Arjoon, S., and Rambocas, M. (2016). Aligning corporate social responsibility with green economy development pathways in developing countries. *Sustainable Development*, 14(4), 237–253.

Shah, N. and Soomro, B. A. (2017). Investigating entrepreneurial intention among public sector university students of Pakistan. *Education and Training*, 59(7/8), 841–855.

Sharma, R. and Rana, G. (2021). An overview of entrepreneurship development programs in India. *International Journal of Environmental Policy and Decision Making*, 3(1), 16. https://doi.org/10.1504/ijepdm.2021.119511.

Shrivastava, P. and Hart, S. (1994). Greening organizations – 2000. *International Journal of Public Administration*, 17(34):607–35.

Soomro, B. A., Ghumro, I. A., and Shah, N. (2019). Green entrepreneurship inclination among the younger generation: An avenue towards a green economy. *Sustainable Development*, 1–10. https://doi.org/10.1002/sd.2010.

Souitaris, V., Zerbinati, S., and Al-Laham, A. (2007). Do entrepreneurship programmes raise entrepreneurial intention of science and engineering students? The effect of learning, inspiration and resources. *Journal of Business Venturing*, 22(4), 566–591.

Speer, T. (1997). Growing the green market. *American Demographics*, 19(8), 45–50.

Sterling, S. (2001). *Sustainable Education*. UK: Green Books Ltd.

Sunny, S. A. and Shu, C. (2017). Investments, incentives and innovation: geographical clustering dynamics as drivers of sustainable entrepreneurship. *Small Business Economics*. Vol(52), 905–927. https://doi.org/10.1007/s11187-017-9941-z.

Thornton, E. (1999). Enviro-cars: The race is on. *BusWeek*, 8(February), 74–5.

Urban, G and Hauser, JR. *Design and Marketing of New Products*. Englewood Cliffs, NJ: Prentice-Hall, 1980.

Vatansever, C. and Arun, K. (2016). What color is the green entrepreneurship in Turkey? *Journal of Entrepreneurship in Emerging Economies*, 8(1), 25–44.

Walley, N. and Whitehead, B. (1994). It's not easy being green. *Harvard Business Review*, (May–June), 46–52.

Wals, A. (2009). *Review of Contexts and Structures for Education for Sustainable Development 2009: Learning for a Sustainable World.* United Nations Decade of Education for Sustainable.

WCED, S. W. S. (1987). World commission on environment and development. *Our Common Future*, 17(1), 1–91.

Wind, Y. J. *Product, Policy: Concepts, Methods, and Strategy*. Boston, MA: Addison-Wesley, 1982.

Wong, V., Turner, W., and Stoneman, P. (1996). Marketing strategies and market prospects for environmentally-friendly consumer products. *British Journal of Management*, 7, 263–81.

Zhaojun, Y., Jun, S., Yali, Z., and Ying, W. (2017). Green, green, it's green: A triad model of technology, culture, and innovation for corporate sustainability. *Sustainability*, 9, 1–23.

5 Entrepreneurship and Digital Economy

Esther K. Mwiti
Technical University of Kenya (TUK)

CONTENTS

5.1 INTRODUCTION

Globalization and modernization have affected the traditional methods of conducting business and there has been a sharp shift to the new era of business operations that is highly automated. Therefore, the policymakers are the watchdog of the economy

DOI: 10.1201/9781003217084-5

hence, need to create an enabling environment for easy adoption of the digital economy. This could be through subsidizing digital resources, adjusting tax rates, and legislation of favorable legal frameworks to encourage more uptake of modern technology by SMEs through the reduction of cost of operations. Designing an economy, which includes all stakeholders, will help to have a long-term sustainable digital economy the legal frameworks of a country aligned with. International frameworks that promote financial and innovation growth will play a key role in enhancing a conducive environment for enterprises. This is echoed in the United Nations report (2019) on the age of digital interdependence.

On the other hand, the private sector needs to partner with the policymakers in order to make it easy for the implementation of interventions made by the government. This will help in the dissemination of the relevant information pertaining to the availability, cost, and use of digital platforms available in the market. A good framework of digital information will give a blueprint of how sound social and economic plans will facilitate entrepreneurs to utilize secure online information in the delivery of services. This is in the World Development Report (2021).

5.1.1 The Genesis of Digital Economy

The digital economy originated in Japan East Asia. According to Thomas Mesenbourg (2001), a Japanese scholar in economics by name of Richard Koo conducted a study on the Japanese economy in the middle of the 1990s. The country's economy was going through a challenging time which he referred to as a "balance sheet recession." This was a result of rules and regulations related to monetary policies that were enacted by the Central Bank of Japan. There was also the world cold war that made the Japanese entrepreneurs experience difficult times. The Japanese economy had also been affected by inflation in the 1980s. Some of the repercussions include:

- A fall of the Japanese land economy.
- Japanese stock market prices experienced a shock.
- Many Japanese enterprises were unable to pay the debts they owed the banks.
- Most Japanese entrepreneurs adopted the culture of saving over borrowing.
- Gross domestic product (GDP) decreased by 1.7%. Many people lost their jobs thus increasing unemployment by 10.8%.
- Many companies went bankrupt due to a long period of a stagnated economy.

The government then reacted by enacting new measures to counter the inflation.

The next time the digital economy was mentioned was in 1995 by Don Tapscott in his book *The Digital Economy: Promise and Peril in the Age of Networked Intelligence*. There were only a few who had heard about the modem. Not even cellphones existed. This was the onset of the digital economy as we know it today. He predicted what would take place two decades following his write-up and today we witness the great evolution of this type of economy. According to Goldfarb and Tucker (2019), digital economics explores how digital technology contributes to economic activities.

5.1.1.1 Understanding of Digital Economy

To an ordinary man, the digital economy is the kind of economy that is connected to the use of the internet and online platforms just to mention a few because it is in a wide spectrum ranging from computer technology, web information, Google platform, and other related information and communication technology (ICT) activities. According to Asian Development Bank (2018), the digital economy refers to a broad range of economic activities that use digitized information and knowledge as key factors of production. This is a new way of carrying out entrepreneurial activities totally different from the traditional way of conducting business. It has been highly influenced by globalization and modernization. Borrowing from Mesenbourg (2001) and Ravindra et al. (2021), we find that today, the world is like a global village since millions and billions of entrepreneurs use online connections and devices to process data, do business, and other service processes in an interconnected manner.

Usage of ICT skills has continued to increase globally reasons being the desire to boost productivity and the current incremental pace of economic and social advancement. It is important to note that the youth population is expanding in most continents of the world, especially in Africa. A recent survey by the International Labor Organization (2018) indicates that a majority of the youths (94.9%) are informally employed. Most of the African countries have faced unemployment and digital technologies offered a chance for employment through various types of virtual opportunities. It is also unlocking new pathways for innovation and creativity for a majority of this age category. To become future innovators, leaders, and entrepreneurs, youths all over the world need to be empowered with digital skills. Access to technology and digitalized markets is essential to thrive in an increasingly digitized global economy. There is a growing need for governments and entrepreneurs to find an effective means of delivering services to citizens, especially in the developing economies. A digitalized economy involves using software & hardware, online transaction, creating a personal brand, or being visible in digital platforms (Rana et al., 2021). So, the digital economy is conducting business activities enabled by the web. The radical changes brought about by digital technology are touching and transforming many peoples' way of living. The main areas of transformation affected either positively or negatively are the entertainment industry, buying & selling, money evolution, photographing, which today is referred to as "selfie" especially when you take a photo of yourself by using a cellphone, schooling & learning, online conversations via WhatsApp, YouTube, Telegram, Twitter, Instagram, among others, and conferencing & webinars. However, access to the internet remains out of reach for most people. For example, in Africa, in 2017 only 22% of the population could access the internet, whereas, in 2020, the number rose to 77%. Many large and emerging economies such as Russia, China, Brazil, Turkey among others are all above 60% in internet use (Jacob Poushter, 2016).

5.1.2 Journey of Traditional Economy to Digital Economy

A traditional economy is a system that is based on basic indigenous activities of the society. A community produces products that they require for indigenous activities common to them. These activities include hunting, gathering wild fruits, fishing,

TABLE 5.1
Traditional Economy versus Digital Economy

Indigenous Economy	Digital Economy (ICT Related)
Physical industries	ICT platforms
Physical business premises	Buying and selling through online platforms
Advertising and promotion by use of sales personnel	Online advertisement
Commuting to workplace	Virtual workers
Physical classes and face-to-face learning	Virtual/online learning
Tangible money (Fiat)	Use of cashless payments such as Mpesa, Cryptocurrency, PayPal, and MasterCard
Security guards/personnel	CCTV and alarm gadgets for alert
Manual recording keeping of sales	Electronic online recording, for example, CDs, related software programs, and most recent cloud computing

Source: Author's own.

subsistence farming, and livestock keeping. Most of the economic activities are founded on the societal values, beliefs, and lifestyles of each community. Brianna Whiting (2021) argues that one of the traditional economic variables is lack of surplus since most of the goods and services produced are fully used. However, the traditional economy has made advance changes due to technological advancement and globalization that have affected ways of living.

On the other hand, Andrew Williams, borrowing from the European Commission (2015), looks at digital entrepreneurship as the broad means of creating new ventures and transforming existing businesses by developing novel technologies and/or novel usage of digital technologies.

Table 5.1 below gives a brief comparison between traditional and digital economies.

Traditionally, businesses operated in a manner where technology was a far-fetched idea. The only technology that was appropriate was needed for businesses to thrive. Most enterprises begin the journey of technology by the application of Appropriate Technology (AT), which is the technology that suits the owner. AT is suitable for every small business and is mainly adopted for local situations. Characteristics of this type of technology are:

- Technology that is simple to use in terms of skills
- Technology that is flexible or that can be easily adjusted
- Technology that is easy to access and acquire
- Technology that is not expensive (low cost or cheap)
- Technology that facilitates access to spare parts locally for maintenance and repair
- Technology that adds positive value to the enterprise
- Technology that can be easily shared out to other businesses in times of need

5.1.2.1 Digital Entrepreneurship

It is commonly said that a journey of a thousand miles begins with a single step. For the digital economy to become a reality, a lot of effort and hard work has been put in place

The digital economy has created a breed of entrepreneurs known as "digital entrepreneurs". This is a kind of a person who is able to gather resources and weigh the risk propensity by assessing environmental conditions in order to offer goods and services for value. They possess substitute traits for a conventional entrepreneur. These attributes are shown in Table 5.2.

In order for a digital entrepreneur to survive in the economy, he/she must keep on coming up with ways of disruption. This is possible by finding new ways of upsetting the market and creating necessary competition, coming up with new rules of the game in order to realize the profit. Many enterprises in various sectors of the economy have succeeded to this end by applying this strategy or taking advantage of disruption innovation (Clayton Christensen, 1995; Prakash et al., 2021). For example, the traditional way of transportation by use of public means or "taxi" are almost falling out of business due to the introduction of Uber as a means of transport. Each sector is reinventing innovative ways to avoid the cumbersome and bureaucratic traditional ways of operation. For this reason, a digital entrepreneur must exhibit very strong characteristics as earlier alluded.

However, not every entrepreneur can be considered to be a digital entrepreneur although they are in possession of smartphones, websites, or doing one or two things on the internet or online social platforms. A digital entrepreneur must possess a mindset of assimilation and inculcate the culture of the digital economy within his enterprise. Most activities/tasks ought to be highly automated rather than apply the indigenous techniques.

TABLE 5.2
Features of Conventional and Digital Entrepreneurs

Conventional Entrepreneur	Digital Entrepreneur
Visionary	Planning
A risk-taker	Effective communication
Perseverance	Adaptability
Innovative/creativity	Knowledge-based
Futuristic	New opportunities
Hardworking and diligent	Tech-savvy (Acumen)
Goal oriented	Hind sight – able to take a reflection on what has already happened
Persistence	Pure determination (requires grit)
Disciplined	Motivated (drive)
Profit oriented	Explorer (adventures)

Source: Author's own.

5.1.2.2 Various Types of Digital Entrepreneurs' Excelling in the Digital Economy

Distinguishing between different types of digital entrepreneurs helps you to know the kind of skill that you require in running your business. This is because not all types of businesses require the same kind of skills or expertise. The following are some of the entrepreneur types that one can choose from:

- **Serial Entrepreneurs:** This type of entrepreneur is one who continuously comes up with business ideas and starts technology-related businesses.
- **Economies of Scale Entrepreneurs:** Transacting large volumes-related type of business in order to cut down the overheads. For example, a service provider in the telecommunication industry.
- **Self-Employed:** Starting and running your own business using digital platforms.
- **Intrapreneurs:** Working for an established organization but exhibiting the characteristics of an entrepreneur on online companies.
- **Opportunistic Entrepreneurs:** Always searching on the internet to find and fill gaps.
- **Speculators:** Take advantage of a situation and invest with a future mind to make a profit on an increased amount of value. They are highly seasonal and may move across industries.
- **Backyard Entrepreneurs:** These operate their businesses from home, either from the kitchen or a home-based office. They use family members to assist in running the business as well use the technology available in their home to create awareness among their customers.
- **Social Entrepreneurs:** This is an entrepreneur who develops a solution to problems within a society by carrying out social projects within his locality.

5.1.3 Technology Uptake Process

5.1.3.1 Planning

For any country to thrive in digital economy, there is a need to envisage a community where the citizens have been enabled to shift from doing all things in a manual way to a digitalized way. It begins by planning and drawing a blue print that will act as a framework for adoption. The blue print is the documentation that bears the vision, mission, and policy guidelines on how the shift should take place. A written document is useful for reference and uses during the stage-by-stage process of setting up structures, procedures, and operations after adoption. It also helps spell out the requirements and outline the nitty-gritty of the process.

5.1.3.2 Conducting Survey

Another important step in the initial stage is to carry out a visibility study of the entire process to find out whether the process is viable or not. The process should include all key stakeholders for ownership so that it is easy at the point of implementation. Training assessment needs are identified through continuous research and development in order to draw training programs for training, the consumers of technology.

5.1.3.3 Training

The training process is vital for consumers like entrepreneurs in order to appreciate that there is a need for a shift from the familiarity of a traditional way of conducting business to a more changed way, where most activities are highly automated. There has been a gradual change in the process of adoption. Most countries are still dragging in embracing technology, especially the underdeveloped and the developing countries. However, there has not been a remarkable change in most sectors of the economy been triggered by natural occurrences like the COVID-19 pandemic. Some entrepreneurs adopt technology not because they are willing but because circumstances push them to do so.

5.1.3.4 Adoption

Adoption of the digital economy is not a one-time activity but a process that requires preparation by the policymakers and the users. The early adapters benefit from high premium rewards that are made by involving in high-risk ventures. For instance, the entrepreneurs who embraced technology and grabbed opportunities in the service-providing industry have graduated into major techno-enterprises. The laggards face stiff competition from already established enterprises that have positioned themselves as the giant firms in these tech industries. The greatest beneficiaries are entrepreneurs that position themselves as service providers. In Kenya, for example, entrepreneurs who are the early adapters in telecommunication such as Safaricom, Telkom, Airtel, among others, have made a kill. As the saying goes in entrepreneurship, "opportunities are for the prepared", many people are risk-averse. Technological businesses require huge capital at start-up stage. For those growing their businesses, or expanding, the cost involved is mainly in upgrading, repairing, management, and maintenance due to the obsolete nature of technology. Enterprises, from time to time, keep on replacing the equipment and incurring costs on capacity building for the needed skills in order to accrue a high return on investment.

Traditional economy mainly involved tangible ways of doing things which were highly manual. For example, being physically in the offices, in the business premises, in the industries, paperwork, involving sales personnel for promotion, tangible money, manual recording & book-keeping of sales, and other types of activities involved in businesses. On the other hand, digitized economy involves information centers, internet space, monitoring business by a click of a button from wherever you are, use of CCTVs, online advertisements, and buying and selling through online platforms.

5.1.4 Major Features of Digital Economy

It is paramount for every entrepreneur who wishes to adopt the digital economy to understand the following characteristics:

Digitized: This means that systems are set to easily monitor and evaluate the processes to acquire data that can be scrutinized to inform at various points and receive feedback for efficient decision making.

Connected: Wireless communications have connected assets, employees, suppliers, customers, shareholders, and other stakeholders. This is beneficial for collaboration, increased efficiency, and responsiveness. It also promotes interrelations among businesses for the purposes of benchmarking, sourcing supplies, connecting with customers, and even subcontracting.

Personalized: Customer specifications in terms of quality products, after-sale service, and quick delivery is key. Thus, optimizing customers' satisfaction which results in more sales and business growth.

Direct: This is the elimination of middlemen and uses a zero-step channel in reaching the consumer. There is also the leveraging of monitoring and evaluation, as well as how to get quick feedback.

Sharing: This is a way of conducting business in which common facilities are shared among entrepreneurs, mainly in the form of assets that maybe lying idle. For example, the concept of Techno Park, business incubation centers, and techno-cities. Various SMEs are clustered together in one area and share facilities.

5.1.5 Components of Digital Economy

According to Mutula (2010), the digital economy consists of several components. The main ones are government policy and regulation (ICT legal framework), internet, electricity infrastructure, e-marketing, e-trading, service providers, and World Wide Web (www), among others. Each of these components plays various roles and functions. Digital economic concepts are philosophically underpinned by the worldwide connectivity of the people, institutions, and objects (facilities) that conduct the processes of technology. This is done through online space and portable technologies that share information from one sector of the economy to another, enhancing the productivity rate. There is also the spillover effect of technology from multinationals to small-scale industries. This happens through the backward and forward relationships, or linkages, that take place between the two. Sussan and Acs (2017) provide a platform to enable us understand entrepreneurship in the current virtual era.

Table 5.3 summarizes some of the components of the digital economy and the role played by each of them.

5.1.6 Merits of Digitalized Economy to Entrepreneurs

The world has witnessed the overwhelming state of an economy that is growing rapidly, bringing into play many innovations that are competing for the resources. This type of economy is not land-locked and therefore can be imported and implicated in far-flung places. Similar experiences can be traced among entrepreneurs globally. The beauty of this type of economy is that you can borrow business models and strategies from other entrepreneurs and other continents and apply them to your business successfully. Since this type of economy is highly collaborative, you can learn new ways of doing things from others who have been able to acquire and use the right skills. This economy has the capacity to build a clientele from all over the world, so the target market is not restricted to the place in which you are operating. The digital

TABLE 5.3
The Components of Digital Economy

Component	Role
Government	• Visionary leadership • Funding infrastructure • Sustainability (maintenance and expansion) • ICT policy regulation • Create demand and supply • Leveraging ICT benefits • Provide a blueprint for adoption • Championing ICT preparedness • Flag out projects • Provide incentives – tax reduction and subsides • Duty-free policy on ICT (hardware)
Policy and regulation (ICT legal framework)	• Enabling environment • Free and fair commerce • Free market economy • National and trans-border transactions • Consumer protection/security • Accessibility and transparency • Competitive business framework
e-Business	• Conducting of business • Organizational processes on computer-mediated networks
Servers and physical systems	• Electronic information • Tangible computer components • Internet connectivity • Intangible assets such as skills, education, training, health, and intelligence of the human resource
e-Commerce	• Online buying and selling of goods and services • Information transmission (data) • Funds transmission (local fiat, forex, and cryptocurrency)

Source: Author's own.

economy is helping entrepreneurs to be more effective and innovate in order to keep abreast with the changing times. Of paramount importance is that once you set your mind, there is no turning back. An entrepreneur is able to explore the myriad opportunities that present themselves in the web space as follows:

- Improves the company's competitiveness through continuous innovations.
- Costs are lowered since most processes are automated.
- Work/tasks are completed in real time, which helps in meeting deadlines.

Table 5.4 summarizes some advantages and disadvantages of the of digital economy.

TABLE 5.4
Advantages and Disadvantages of Digital Economy

Advantages	Disadvantages
Greater information and choice	Monopoly by large companies
Time-saving	Technology addiction
Less costs	Less community
Greater personalization	Not complying with regulations
Lower barriers to entry	Privacy issues
More flexibility by working from home	Disruption to traditional economy and jobs
Economic developments	Potential environmental costs

Source: Author's own.

- Scheduling work is easier.
- Inspection of work processes is possible throughout the process.
- Benefits are shared across the supply chain.
- More quality products are produced, enhancing productivity rate.
- Monitoring and evaluation are easier with the use of surveillance and other relevant systems.
- Increased market opportunities since you can sell across the world.
- Enhance foreign exchange.
- Generate higher revenue for the government.
- Higher percentage of GDP.
- The interconnectedness of larger and small enterprises through backward and forward linkages.

5.1.7 Main Pillars of Digital Economy

A pillar holds together the framework of any structure. The digital economy is structured in a segment of its kind that is surrounded by several blocks of technology. Some of these include cyberspace, online correspondence, online payments, computerization, and the internet community.

Cyberspace: This is the connectivity to the www that helps entrepreneurs search for information related to their products and services. In this space, you can find all types of information that you need. Entrepreneurs can also avail their products' information through the same space with an intention of creating awareness among the consumers who may be searching for their goods and services.

Online Correspondence: This is a method of exchanging messages between people using electronic devices. Entrepreneurs can use this method of exchange to pass information because it has wonderful features such as efficiency, speed, reliability and can also cover long distances.

Online Payments: These are payment methods that are contactless, cashless, and paperless. Traditionally, entrepreneurs exchanged their goods using their local currency, but today, world-wide currency and payment methods have been invented. Some of these payments are local innovative ways of payment specific to a country where they have been developed, while others are global.

Computerization: Entrepreneurial operations and processes are empowered by technologies rather than operating manually by high use of human labor. The performance of tasks by machines results in quality products, efficient and timely. By use of these techniques, entrepreneurs can perform a wide range of tasks across all sectors.

Internet Community: These are entrepreneurial web families who are attracted each other with the help of technology. They support each other by sharing enterprise information and commend on entrepreneurial matters. These families share information about new trends in the market, availability of products, ways of advertising, and promotion among other things.

Examples of the digital economy include:

i. In the transport industry, Uber, among others, has greatly helped in the transformation of the industry.
ii. The hotel industry is also not lagging behind. Platforms such as Booking.com, Expedia.com, and Hotelonline.com in Kenya, among others, have made travel and accommodation bookings easier.
iii. Online marketplaces – this is where entrepreneurs buy and sell their products electronically. It is a channel that is used to avail information to those consumers searching for particular products or services in any part of the world. It has an advantage in that you have a wider target market that is able to reach to all segments of the clientele by a click of a button.

5.1.8 Transforming Entrepreneurship through Technology Transfer

Innovation, creativity, and disruption are great cornerstones of the digitalized economy. The reasons are that entrepreneurs should be keen on the changing business terrain or landscape. The changes that are occurring are not for a short time but are likely to linger for a while or completely change the landscape of businesses. Training for skills and retention of already built abilities and talents should be a business model and strategy that should not be abandoned. Every business person needs to know that it's not a 1-day activity but a process that involves technological transfer for a long period of time. This technological transfer shown in Figure 5.1 begins with a comprehensive survey in order to identify the available technology and evaluate its suitability or appropriateness to the enterprise requirements. Not every technology is fit for use by every enterprise. It is dependent on the benefits the entrepreneur wishes to draw from it. A criterion for accessing the best technology is a key step in making the right choice of technology that you would wish to adopt or absorb in the enterprise. However, several companies may share technology by replicating from one business to another. Figure 5.1 shows the step-by-step procedure for technology transfer.

After the entrepreneur has adopted the AT in the enterprise, there is a need to automate most of the processes for maximum returns or for a higher productivity rate. Generally, entrepreneurs hate routine. They get bored easily and therefore repetitive tasks that can be automated should be channeled to machines to cut down on

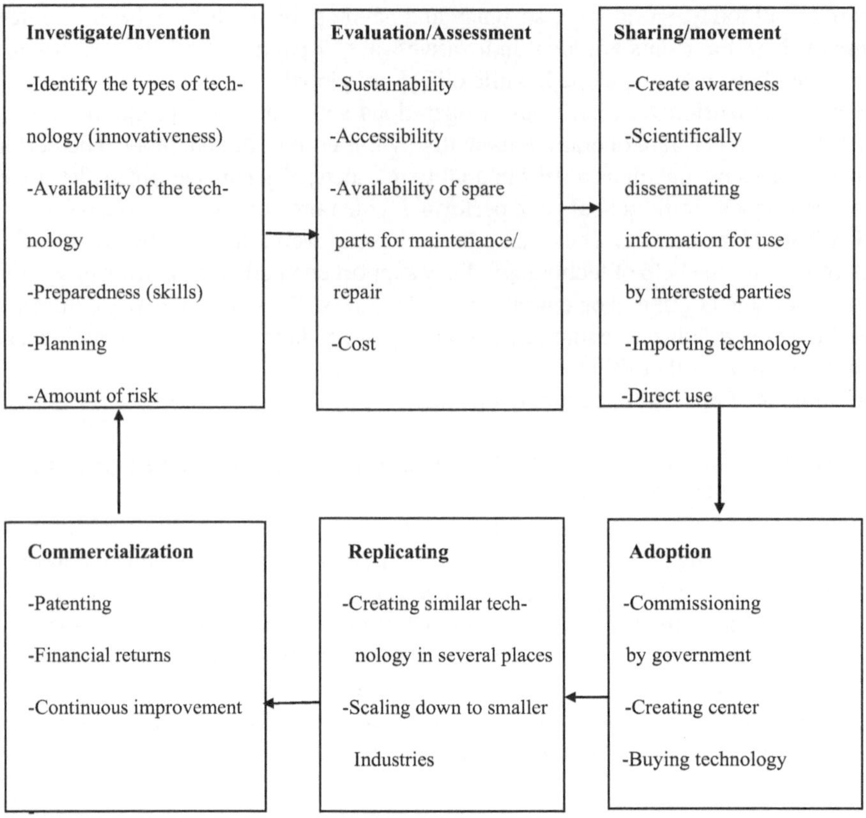

FIGURE 5.1 Step-by-step procedure for technology transfer.

production costs and increase efficiency. This is a reason why the digital economy is more preferred by entrepreneurs than the traditional ways of trading.

5.1.9 Entrepreneurial Creativity and Innovation in Digital Economy

Primarily, the digital economy involves a lot of creativity and innovations. Entrepreneurs become more creative when they are motivated by their own interests to satisfy needs and provide solutions to challenges. Motivations are not determined by external pressures but also internal factors like passion and interest that drives you to desire to do something unique. Creativity is rethinking how to do new things and come up with new ideas through your own imagination (Rana et al., 2019). You figure out how to provide a solution to problems and by making use of available resources, you pursue an opportunity. Innovation, on the other hand, is the ability to utilize new methods of doing things, new processes, and new ways in order to enhance people's lives or to uplift the standards of living in a society. Creativity is the prerequisite of innovation meaning innovation starts with creating ideas. One is inspired to create things differently from others by cutting out a niche in the marketplace. This helps an entrepreneur differentiate themselves and make an impact. To excel in creativity, you must become an expert whereby you encompass anything that you can do in the broad domain of your work. One such is having knowledge and technical ability. Other components of creativity involve creative thinking and drive & desire to do something. The digital economy is highly powered by technology. Innovative technology involves creative thinking processes that generate solutions by the force of imagination and reasoning. Technology is a product of thinking and is manifested when the thinker's mind is exposed. People who see problems as opportunities to improve and do something new can improve their entrepreneurial world by diffusing technology in their enterprises. In this case, this entrepreneur will be asking "how can I go about it?" or "what do I need to do to make things better?"

5.1.9.1 The Creative Process

There are a few steps that complete the process of creative thinking. It begins by building information, incubation, idea, implementation, and evaluation.

1. **Building Information:** Information is built by searching widely. Searching can involve extensive reading, participating in conferences, webinars, professional meetings, and engaging experts like consultants in conversations. Once these data are processed, it becomes knowledge. When you accumulate knowledge, you have a chance to progress in life. An entrepreneur searching for opportunities in the digital economy should read widely and acquire knowledge related to his area of interest.
2. **Incubation:** This is where you allow ideas to take shape and come out full-blown while you are engaged in activities totally unrelated to the subject of concern. You allow your mind only to think about the problem you want to solve from time to time. For example, before falling asleep or meditating on it first thing when you wake up, or while going on a walk or some kind of exercise. You can also observe what others are doing.
3. **Idea:** In the creative process, this will be said to be a hint in your mind related to a solution to the problem. You can daydream, or record anything that is crisscrossing your mind as you engage in a moment of leisure or break times from your normal activities.
4. **Implementing:** This is putting the idea you have into action. One must be knowledgeable and energetic to carry out the implementation process. You can consult with experts in order to test your idea with knowledgeable people. Whenever possible have a plan that is drawn to carry out the process.
5. **Evaluation and Monitoring:** this involves the assessment of the process of implementation, finding out the success point and failure point of objectives and goals. You look at the deviations so that you can make adjustments (Rana and Sharma, 2019). It also means identifying the risks that you are likely to encounter in a broader term.

5.1.9.2 Innovation Process

The innovation process involves analytical planning, resource mobilization, applying resources to accomplish plans, and commercialization. Digital transformation is a business model that has dynamic potential to change entrepreneurship across all sectors which are applying innovative policies. Recently, there are development toward more service-oriented industries than only dwelling in manufacturing. A company can choose to develop both software and hardware products. This is a call to policymakers and enforcers to include both manufacturing and service industry policies for technological companies. According to Organization for Economic Co-operation and Development (OECD, 2017), enterprises are shifting toward information-driven innovative strategies. Secondly, there is more flexibility in order to increase speed of innovation.

Therefore, entrepreneurs in a digital economy require innovation and creativity wisdom in order to take advantage of:

- New skills (IT-related talents and abilities).
- Entrepreneurial competencies (time management, decision making, change management, goals setting, and risk management).
- Multidisciplinary investigations (research and development in various disciplines).
- Be able to create entrepreneurial teams whose composition is drawn from experts in various fields.

The above factors will help an entrepreneur to leverage and thrive in the digital economy.

5.1.10 Maximize Technology for Higher Productivity

We begin by asking ourselves if has technology improved productivity or has it caused its decline?

In order to achieve full productivity with the use of technology, entrepreneurs should consider the following:

Use of the AT: Not every technology is fit for use for every business. Some of the enterprises may require the basic minimal or the bare minimum. Indeed, some can even improvise technology, whereby they come up with innovative ideas and concepts on how to go about things. This type of technology can be the cornerstone to building on or conducting further research on better tech for their businesses. Improvising is creativity in expression and application on a smaller scale, which is essential or necessary for a productive enterprise. Creating a conducive environment for technology movement in an enterprise will provide an impetus that will propel technological growth toward a more digitalized economy.

The Cost of Technology: Entrepreneurs should make use of technology that is convenient, available, and affordable. Not every activity in the enterprise will require the use of technology at all times. Sometimes you can avoid the use of technology on tasks that can be easily performed manually. An entrepreneur should highly avoid the misuse of technology. You don't have to have the computer think for you all the time one needs to figure out how things should be done and arrive at results without necessarily having to use technology. This will highly reduce overhead costs.

Tech-Integrity: This is a digital value system that enhances strong moral principles in dealing with clients. Sometimes the privacy and security of the enterprise can be compromised by the use of technology. In delivering quality goods and services, the entrepreneur will need to guard against dishonest deals, "the get rich quick deals" that usually promote cyber theft, compromising the business integrity. Much as technology is beneficial in business operations, it can bring along with its cyber-crime that requires cyber security for survival. Business strategic objectives can be highly achieved and on record time by many entrepreneurs but some are apprehensive due to fear of exposure. Lack of trust is one other reason why many people are not connected. The enterprise owners must make sure that their customers are comfortable using technology and can trust them for the safe delivery of goods and services (Singh et al., 2019). There have been cases when entrepreneurs place orders online with correct specifications only to receive products that don't meet quality. This is just an example of many cyber-crimes that takes place every day. So, an entrepreneur who needs to excel in a digitalized economy must be vigilant.

Digital Ethics and Governance: Regulations to deal with tech crimes and misuse. This comes up with requirements and implementation, but putting them to use may be a challenge too. Find out what others are doing, collaborating, learning, and applying. Rules must be crafted in a way that they will not hinder innovativeness in the company or create untold fear or anxiety that can cause workers not to progress. The code of ethics in relation to the use of technology in the enterprise. Rules and regulations are important for helping the human resource in the enterprise to remain within the set boundaries of technology use. Following the right procedures as stipulated in the set work standards will help in maximizing technology on the right tasks with very minimal deviation. For example, in the case of using a smartphone or a laptop, one can easily be carried away to other platforms and lose track of the tasks at hand.

Innovative Use of Technology: Technology has been in use since the evolution of human beings. During the stone-age period, man discovered how to lit the fire & cook, warm himself, and protect himself from animals. He also discovered how curve flint knives are used as weapons for hunting, not to mention innovating ways of taming a dog to help him in hunting. So, every stage of man involves new inventions, discoveries, creations, and innovations (Sharma et al., 2021). These four are the elements of entrepreneurship. For enterprises to be productive in the digitalized economy, these four elements must fall in their place. More technological inventions, discoveries, creations, and innovations keep on fleeting, and those entrepreneurs that lodge in the corridors of technological development grab these endless opportunities.

Continuous Improvement/Training: Entrepreneurs should be proud to start with the already available technology within their reach and keep on improving as tasks demand a more sophisticated technology. This is because technology runs obsolete very fast, therefore creating a demand for a more advanced one, which can be very costly. Having a totally digitalized enterprise may call for advancement in skills, which can be learnt on the job as you perform the tasks and as more machines are introduced.

Embracing Technological Change: In a dynamic world, change is inevitable. Regardless of how much many people resist change, it is unavoidable. Technology has witnessed a lot of changes, from the basic to the more advanced type. Various types of objects in technology have been developed that aid entrepreneurs at various points of the value chain. Technological processes and technological strategies for how to do things have reshaped the economy. The nature of technology has portrayed multifaceted technologies, provoking enterprises that are result-oriented to come up with business models that increase productivity and counter competition while developing the coping mechanisms presented by technology advancement. Hence, developing systems and programs for mitigating changes presented by technology cannot be brushed off by any successful entrepreneur.

Other key requirements an entrepreneur needs to have in the subconscious include:

- Availability of information or data that need to be shared or exchanged
- Digital infrastructure that is effective for use of transmission
- Permission, licensing that is the required authority to operate that kind of economy
- Guidelines and policy documents that give directions
- The ability of computer machines to transmit and use coded data.

5.1.11 Global Pandemic (Covid-19) Reshaping Digital Economy

Life phenomena whether by the natural occurrence or caused by mankind results in changes that often trigger the human mind to provide solutions to their challenges. COVID-19 is one crisis that has affected people around the world. COVID-19, with its associated effects like quarantine and lockdown, is an entrepreneurial disruption that has plunged the world economy significantly. Many entrepreneurs have been forced to make a sudden shift from the traditional way of doing business to adopting the new era way. This paradigm shift was realized by those who were technologically prepared, either by having previously used AT in their enterprises or by having the capacity in terms of computer skills proficiency, and who already owned or were able to acquire needed gadgets such as smartphones or computers, utilities such as WiFi or fiber for connectivity. We need a technopreneur mindset to take up a competitive approach and innovative strategies in order to counter the economic disruptions successfully. In this case, entrepreneurs had to rethink and restrategize on how to plunge into this type of economy without major injuries that could result in the closure of businesses. Some of the innovative possibilities that most entrepreneurs embraced included Zoom, Google Meet, Skype, and other types of social media platforms. Some of these alternative platforms have greatly affected communication and have been widely embraced widely. After physical meetings were banned, entrepreneurs and corporates were able to conduct their meetings virtually. There has been a mushrooming of cottage industries whereby businesses have been conducted in the backyard and comfort of entrepreneurs' homes.

The COVID-19 crisis is provoking the entrepreneurial spirit to focus on more productive contributions to the digital economy. Even so, the global economy is opening up to many people and new opportunities are streaming in every day. The disruption of the economy by COVID-19 brought forth a situation whereby entrepreneurs changed the game rules. Not many people anywhere in the world love to leave the comforts of their beds and wake up to go to work every morning. On the onset of COVID-19, the uncertainties experienced by the enterprises increased in a terrifying way. This led to the diffusion of technology into work systems that presented opportunities for digital entrepreneurs. In Kenya, for example, just like many other countries, I suppose, started to allow minimal numbers of workers to be physically in the office while others worked at the comfort of their homes. Work plans, schedules, and timetables were drawn away from the normal routine to allow people work in shifts. New methods of monitoring and receiving feedback or results became a discovery. Service-providing companies were engaged by most enterprises that operated highly in manual way to train staffs and set up automated systems to assist workflow in the "new normal".

As a result of these notable changes, many entrepreneurs have digitalized their operations, and online sources of business opportunities have increased allowing more jobless people to join the wagon of work through self-employment. Marketing of products through digital platforms and branding of companies expanded. For instance, digital platforms have been used for survival by the creative industry. A case example of Kenya shows that musicians have found new grooves in the time of COVID-19 to supplement their income. This is because the COVID-19 pandemic affected the live performance and productions. When performers are experiencing income pressures, they feel frustrated and production gets affected. Some used their income reserves for sustainability. COVID-19 has seen many creative turn to digital platforms and technologies for new entrepreneurship opportunities.

Some of the platforms that have been useful include;

- Dundalive allows e-payments for streamed content as well as Tip line through Mpesa mobile money solution through PayPal for those followers outside Kenya.
- Mdundo, which is used to download artist music tunes and ring tones for mobile phones. A report by the United Nations Conference on Trade and Development (UNCTAD, 2020) showed that 22% of 5 million African artists using this platform are Kenyans.

A survey by HEVA Fund (2020), Africa's first dedicated finance, business support, and knowledge facility for creative industries, similarly underscored the financial impact of COVID-19 on creative business in Kenya.

A strand of new entrepreneurship has emerged in the religious sector whereby, as a result of worship centers' lockdown, more innovations have come up in the way spiritual entrepreneurs participate in creating enterprises other than the traditional way of generating revenue. Since some of the companies restructured or cut down the pay cheque, those rendered redundant joined the world of digital entrepreneurship by launching businesses digitally. There has been a mushrooming of mobile grocery businesses, online shopping platforms, advertising of services, school learners' materials & instruction programs, and YouTube learning short courses on how to acquire skills like baking and mat-making in order to develop cottage industries. The COVID-19 pandemic has seen an increase in the number of people participating in the businesses related to market networking in companies that supply nutritional supplement products. There has been a high level of acceptance of new e-money that is traded by use of technology like block chains and online platforms. Echoing the words of Peter Beech (2020), "COVID-19 may be having a devastating impact on our industries, social lives, and personal grooming standards, but it is also prompting an outpouring of creativity in other arenas." Even if this has been a challenging moment, the COVID-19 pandemic has enabled major discoveries, inventions, innovations, and creativity in relation to entrepreneurship.

5.1.12 Emerging Trends in Digital Economy

5.1.12.1 E-Money Platforms

There are several digital assets that have been used in the evolution of e-money. A digital asset can be defined as information that is stored digitally and has value and ability to be used. This is a new concept that developed in the mid-90s where items such audios, files, videos, spreadsheets, among others, came into play. Today, digital assets incorporate websites, content management mainly for marketing, social networks, and high-end digital assets such cryptocurrencies. These digital assets have increased wealth creation and numerous opportunities for investment, especially during the COVID-19 pandemic. E-money technologies have come into play in the recent past. One such is block chain technology, which is a decentralized, digital ledger that acts as a database that helps store transactions in a manner that no alterations or delete is possible by anyone. The block chain has several blocks that store digital pieces of information.

5.1.12.2 Cryptocurrency

Cryptocurrencies are digital or virtual currencies that are well secured to avoid counterfeiting. Many entrepreneurs are finding opportunities in this type of currency, though it is risk-intensive. There are several types of cryptocurrencies, just to mention a few: Bitcoin, Ethereum, Tether (USDT), Lite coin, and Doge coin. Bitcoin is the highest in value.

It is advised that digital entrepreneurs read the disclaimers of each company that is trading in digital currencies and also consult their financial advisors before undertaking this type of venture. There is an assumption that this type of e-money will be the way to go in future since many entrepreneurs are beginning to appreciate the potential in this industry. Some are proposing that we are headed towards a single currency economy through e-money.

5.2 CONCLUSION

The digital economy is taking up the world's space as an innovation that is inevitable. Every entrepreneur that is passionate about leaving a mark and legacy in the business fraternity must be ready to compete with the world space of technology, or rather, cut out their space. Capacity building in the area of ICT is important even for the marginalized and those who have not yet embraced modern technology to allow for preparedness. The digital economy is here to stay, and therefore, everyone should embrace an entrepreneurial mindset and acquire the necessary capability for a transition or a shift from traditional economy to this type of economy.

REFERENCES

Adhiraj, S., (2015). *The Importance of Technology in Entrepreneurship.* EPI Centre Ambassador. University of Windsor Campus. Retrieved from http://www.epicentreu-windsor.ca.

Antonio, T., (2019). Developing entrepreneurship in digital economy. DOI: 10.5772/Intechopen.

Asian Development Bank. (2018). *Understanding the Digital Economy: What Is It and How Can It Transform Asia?* New Delhi: ADA Institute.

Brianna, W., (2021). What Is a Traditional Economy? – Definition, Characteristics, Advantages & Examples.

Clayton Christensen, M., (2015). Strategy & innovation book. *Harvard Business Review.* 24(9).

Don, T., (1995). *The Digital Economy: Promise and Peril in the Age of Networked Intelligence* (1st ed.). McGraw Hill.

Flavio, C., Chiara, C., Rudy, V., (2020). *Start-Ups in the Time of COVID-19: Facing the Challenges, Seizing the Opportunities.*

Goodman, P., (2021). *The Pros and Cons of Technology.*

Granta, P., (2021). *Entrepreneurial Models for Industry 4.0.* Great Abington, Cambridge.

HEVA Fund, (2020). *COVID-19 Resilience: Creative Industry Options and Strategies.* Nairobi.

JadMawlaw, (2020). *Leading the Organization in Global Business Strategy, Media and International Relations.* CEO at Dooply Headquartered in the United Kingdom.

Mutula, S., (2010). *Digital Economies: SMEs and E-Readiness,* University of Botswana.

Najwan, (2019). *Traditional VS Digital Business Practices from I Price Group 3rd February.*

OECD, (2017). *Innovation and Digital Economy, in OECD Science, Technology and Innovation Outlook 2017.* OECD Publishing, Paris.

Peter, B., (2020). *These New Gadgets Were Designed to Fight Covid-19.* World Economic Forum.

Poushter, J., (2016). *Smartphone Ownership and Internet Usage Continuous to climbIn Emerging Economies.* Pew Research Center, Washington, DC.

Prakash, C., Saini, R., Sharma, R. (2021). Role of internet of things (IoT) in sustaining disruptive businesses. In R. Sharma, R. Saini & C. Prakash, V. Prashad, *Internet of Things and Businesses in a Disruptive Economy* (1st ed.). Nova Science Publishers, New York.

Rana, G., Sharma, R. (2019). Emerging human resource management practices in Industry 4.0. *Strategic HR Review,* 18(4):176–181.

Rana, G., Sharma, R., Goel, A. K. (2019). Unraveling the power of talent analytics: implications for enhancing business performance. In: Rajagopal, B R. (eds). *Business Governance and Society.* Palgrave Macmillan, Cham. DOI: 10.1007/978-3-319-94613-9_3.

Rebecca, M. (2018). *Consider the Differences between Traditional and Digital Businesses.* Published under Digital Transformation.

Sharma, R., Rana, G., Agarwal, S. (2021). Techno innovative tools for employer branding in industry 4.0. In: G. Rana, S. Agarwal and R. Sharma, (ed.), *Employer Branding for Competitive Advantage* (1st ed). CRC Press, Boca Raton, FL. DOI: 10.4324/9781003127826-11.

Sharma, R., Saini, R., Prakash, C., Prasad, V. (2021). *Internet of Things and Businesses in a Disruptive Economy* (1st ed). Nova Science Publishers, New York.

Singh, P. J. (2019). *India Should Aim for a Digital Non-Alignment.* IT for Change, Bangalore.

Singh, R., Anita, G., Capoor, S., Rana, G., Sharma, R., Agarwal, S. (2019) Internet of things enabled robot based smart room automation and localization system. In: Balas V., Solanki V., Kumar R., Khari M. (eds). *Internet of Things and Big Data Analytics for Smart Generation. Intelligent Systems Reference Library*, vol. 154. Springer, Cham. DOI: 10.1007/978-3-030-04203-5_6.

Spiezia, V., Tscheke, J. (2020). International agreements on cross-border data flows and international Trade: A statistical analysis. *OECD Science, Technology and Industry Working Papers, No. 2020/09*, OECD, Paris.

Srinivas, K. Yasmeen, S., (2017). A study on employee engagement in small and medium enterprises in NBER Macroeconomics Annual 2020, Vol. 35, University of Chicago Press.

Sussan, F., ACS, Z. J. (2017). *The Digital Entrepreneurial Ecosystem. Small Business.*

United Nations (2019). *The Age of Digital Interdependence.* Report of the UN Secretary-General's High-Level Panel on Digital Cooperation, New York.

World Bank, (2021). World Development Report 2021: Data for Better Lives. Doi: 10.1596/978-1-46481600-0. World Bank, Washington, DC.

6 Venture Capital Financing
Methods and Strategies

Jeevesh Sharma
Manipal University

Roomi Rani
Govt. SPMR College of Commerce

Saurabh Joshi
Swami Rama Himalayan University

CONTENTS

DOI: 10.1201/9781003217084-6

6.1 INTRODUCTION

Venture capital is an important source of funding for small- and mid-sized businesses that have few other options for generating capital. Such businesses may have a lot of potential to make a lot of money and grow into larger companies in the future. However, because of the danger inherent in these sorts of investments, most ordinary investors are hesitant to invest their money in them. The notion of venture capital arose to offer financial assistance to such entrepreneurial potential and business abilities.

Entrepreneurial businesses are the underpinning of economies, generating employment, and that results in the expansion of the economy (Sharma & Rana, 2021). Young, creative, and innovative entrepreneurial businesses are essential to the development, expansion, and introduction of new technologies, sectors, and markets. However, to maintain their development, these companies frequently require significant sums of financial resources. Entrepreneurial finance literature has underlined the role of venture capital and business angels in recent decades. The need for funds by entrepreneurial firms or startups has given birth to a new type of financing in the market, that is, venture capital financing, angel investors, etc. (Bonini & Capizzi, 2019). To start their business, entrepreneurs can borrow funds from several sources such as accelerators and incubators, proof-of-concept centers, university-based seed funds, crowdfunding networks, and intellectual-property-backed financial instruments, which are the examples of Intellectual Property (IP) backed financial instruments (Bellavitis et al., 2017). Thus, the globalization of financial markets has made it possible for businesses to raise funds from investors across the world.

Recognizing the centrality of entrepreneurial businesses to the entire economy, more study is required to understand how these various sources of funding affect startups and how these novel events alter current ideas. In the last few years, there is a boom in research on entrepreneurial finance and how the startup gets funding, how they initiate their business based on strategy, and what are the major sources of funding for startups. But, before establishing any business, the person needs funds, and in the case of a startup or an entrepreneurial business, the bank or any financial institution does not provide funds to the entrepreneur because of an absence of historical performance of the business. Due to not having any source of finance for entrepreneurs, the term venture capital was coined, and the idea of venture capital was developed to offer financial assistance to entrepreneurs based on potential and business abilities.

With the burgeoning importance of entrepreneurship and its significant impact on economic development, this book is introduced to enrich the knowledge of academics on entrepreneurship and entrepreneurial financing. This chapter explores the evolution of entrepreneurial finance and venture capital across the market. It also discusses what forms entrepreneurial finance and how venture capital marked its space in the financial market. This chapter also determines the factors that give birth to venture capital and discusses the origin and role of venture capitalist (VC) in the Indian economy, along with major venture capitalists operating across India. The features, methods, and tactics of venture capital financing in the growth and development of startup businesses are explained. Therefore, the chapter provides a snapshot of venture capital across India, its growth in various sectors of business, the process of funding by the venture capitalist in India, and which are the startup firms that are funded by the various venture capitalists.

6.2 ENTREPRENEUR

The word "entrepreneur" is derived from the French word "entreprendre," which means "to undertake," "to pursue chances to fulfill wants and desires through innovation," and "to engage in business." Richard Cantillon, a social scientist, was the first to apply the term "entrepreneur" in economics, in 1725. In 1961, David Mc Cellion developed the philosophy of entrepreneurship.

An entrepreneur is a person who starts a new firm and bears most of the risks while establishing a business. The entrepreneur developed a business model for their business to ascertain the company's core profit-making plan. It describes the product and targeted market of the product. Thus, Business model brings together capital, land, and labor for products or services. The main characteristics of an entrepreneur are risk-taking ability, creativity & innovation, need for achievement, leadership capability, the desire for autonomy, etc.

6.3 ENTREPRENEURSHIP

Entrepreneurship is a process of creating or forming a startup or a new young company by an individual/entrepreneur. Entrepreneurship is an art of creating a company, which is essentially a new company or startup, offering a unique product, procedure, or operation. It is possible to describe it as a creative activity, where an entrepreneur exhibits his idea and converts it into a business. An entrepreneur seeks everything as a chance and is biased when making decisions to take advantage of it. An entrepreneur is an expert in both technical and managerial skills; these skills make an entrepreneur create his own company or startup. Entrepreneurship is an economic and dynamic activity that entails developing and operating a business to optimize the use of limited resources to create value/money. Entrepreneurship is seen as a dynamic force because this value creation activity is carried out constantly in an uncertain business environment. It also involves continuous evaluation of the business model at every stage of a business life cycle to avoid any losses, in short, it means an optimization of performance.

> Entrepreneurship is the pursuit of opportunity beyond resources controlled.
>
> *Stevenson*

Finally, entrepreneurship must be viewed as a means of self-transformation and, as a result, becoming a change agent. To make this happen, the entrepreneur must be able to meet the larger system's environmental and socioeconomic expectations while also driving themselves to achieve their goals. The most essential driver of success is the balance between inner desires and external expectations.

6.4 ENTREPRENEURIAL FINANCE

Any new business needs finance, so entrepreneurs must decide from where to get it, how to invest, and how much to borrow. Entrepreneurs' primary concerns are where and how they will obtain capital to initiate their companies and get them off to a good start. While bank loans and venture capital finances are often linked to entrepreneurs and entrepreneurial companies, some aspects of both the sources are common. Venture capital finance is a major source of providing funding to startup firms, as entrepreneurial firms are usually young and have a high risk of failure. So, both venture and bank loans required the continuous monitoring of firm performance. Both the sources of funding have some terms of the agreement that are used in both types of financing to limit the debtor's conduct and give extra control in the case that the company performs negatively. Thus, these two types of institutions have quite distinct financial systems. Banks provide relatively liquid assets to their customers, exposing them to the risk of unexpected losses. Investors in venture capital funds are subject to limitations, which protect the funds from unexpected losses but force them to pay a premium to prevent loss of liquidity (Winton & Yerramilli, 2008).

Entrepreneurial finance is a third key source of finance in the financial market after personal finance and corporate finance. Entrepreneurial finance is essentially defined by the nature, demands, and dynamics of a new enterprise/venture, as well as the entrepreneurial side. All this comes down to using basic financial ideas and theories in the context of young, innovative, and small-scale enterprises. This also refers to adopting such principles and theories for identifying & designing, launching, managing, growing & maturing, valuing, and harvesting entrepreneurial business enterprises. The confluence of the two areas of "entrepreneurship" and "finance" is referred to as "entrepreneurial finance." Entrepreneurial finance includes venture capital, private equity, private debt, trade credit, Initial Public Offering (IPOs), angel finance, and crowdfunding, among other forms of finance (Cumming & Johan, 2017). The capacity of entrepreneurial businesses to acquire financing is one of the most pressing challenges they face. As a result, venture capital firms, angel investors, and corporate investors are the three major sources of outside equity funding for entrepreneurs. Limited partnerships in which the managing partners invest on

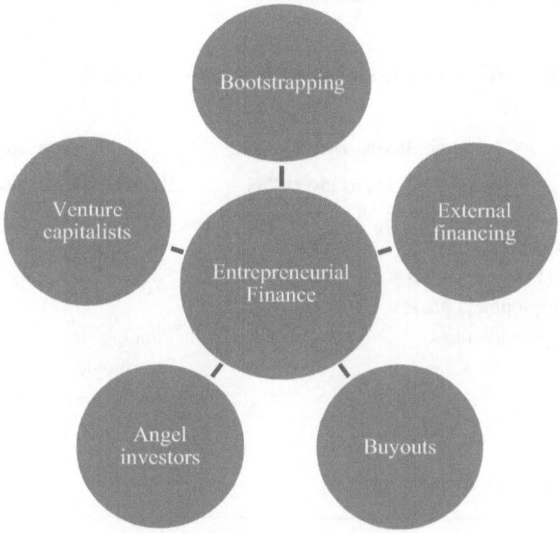

FIGURE 6.1 Sources of entrepreneurial financing.

behalf of the limited partners are referred to as venture capital funds. Angel investors are high-net-worth people who invest their own money in a limited number of businesses. For financial and/or strategic reasons, corporations invest on behalf of their shareholders (Denis, 2004). Figure 6.1 depicts the various sources of entrepreneurial finance. Each source of finance is discussed in the next section.

6.5 BOOTSTRAPPING

In this type of venture financing, the entrepreneur invests his own money in the company and proposes the share to an individual in the exchange for services provided, in addition to another type of financing is also included like deferring the payment of partners and offering equity participation to the employees and stakeholders. The main thing to remember about bootstrapping is that it can only be implemented when the entrepreneur does not require a large amount of capital, as all of the approaches listed above are related to investments with limited capital mobilization.

6.6 EXTERNAL FINANCING

Under bootstrapping financing, an entrepreneur obtains funds from internal sources or by offering shares in exchange for work, whereas, external funding frequently entails obtaining funds from outside sources that are tangible and immediately fundable.

6.7 ANGEL INVESTORS

Most of the time angel investor and venture capital terms are used interchangeably, but there is a minor difference between these two concepts. Angel investors, as previously defined, are high-net-worth individuals who share their wealth in small-scaled

TABLE 6.1
Difference between Angel Investor and Venture Capital

Basis of Difference	Angel Investor	Venture Capital
Background	High-net-worth individuals invest their money in new and small businesses and sometimes former entrepreneur	Form of Private equity, is generally a group of well-off investors, investment banks, and investment institutions.
Investment	Seed stage or initial stage of startup development phases	Seed and later stages also
Deal structure	Angel networks	Funding stages (at every stage of startup development) and funding rounds (A, B, C)
Exit	Active mentorship	IPO, repurchase of shares, etc.

Source: Author's own compilation.

businesses (Chemmanur & Fulghieri, 2014). Furthermore, unlike the venture capital industry angel investments are confidential deals that are not to be disclosed to the public. Angel investors are a group of investors or a network of angels; they do not have any formal institution, but before investing, the angels frequently evaluate all pitches of business plans and conduct due diligence and then invest in some of them. Angel investors tend to invest in the early stages of startup where the company requires a substantial amount of money to trigger a business. Angel investors are those individuals who not only invest in the entrepreneur's business but also provide management expertise and actively mentor the entrepreneur until the business gets established and successful in the market. Moreover, to develop social responsibility and give back to society in the form of funding new ventures is the prime motive of angel investors. Angel investors can be considered as game-changer in the development of startups, as they believe to invest in innovative, young small businesses. Angel investors are complimentary to venture capitalists; angles invest their funds in startups and work with their networks, whereas the venture capitalists provide investments from pooled investment through numerous investors. Table 6.1 summarizes the disparity between angel investors and venture capitalists.

6.8 VENTURE CAPITAL

6.8.1 Meaning and Definition of Venture Capital

The term "venture capital" can be interpreted in various ways. In a narrow sense, it refers to funding for new and untested businesses that are without a consistent record of growth. It refers to an investment in commercial initiatives where uncertainties are yet to be minimized to a risk that can be assessed using sensible security criteria.

Wealthy investors, investment banks, and other financial institutions are the most common sources of venture capital. Not only does it assists in the form of money but also in technical or management expertise. Small businesses with extraordinary development potential, or businesses that have developed fast and look set to expand, are often given venture capital. Venture capital investment is money given to entrepreneurs. It can be supplied at various phases of their development, although it is most commonly associated with early and seed round investment. Venture capital funds are generally only accessible to authorized investors and manage pooled investments in high-growth prospects across startups and other early-stage companies. It can be understood with this also, that large ownership portions of a company are built and sold to a few investors through separate limited partnerships established by venture capital companies in a venture capital deal. Occasionally, these partnerships are made up of a collection of comparable businesses.

Venture capital is a form of private equity, here private equity means when an entrepreneur approaches a venture capital company and an individual, by suggesting them to invest or finance in their business, they would be offering you an equity share in their company. However, one significant distinction between venture capital and other private equity deals is that venture capital focuses on emerging companies looking for major funds for the first time, whereas private equity funds larger, more established companies seeking an equity infusion or the opportunity for company founders to transfer some of their ownership stakes. Management and technical knowledge might be included in venture capital. The majority of venture capital originates from a collection of rich investors, investment banks, and other financial organizations that pool their funds to form joint ventures. This method of generating cash is common among new businesses or enterprises that have little or no operating experience and are unable to secure financing through a debt offering. Thus, VC invests in the idea of entrepreneurs and assists them in establishing business after analyzing the growth potential of business plans and providing both financial and management expertise to startup companies.

6.9 VENTURE CAPITALIST

A venture capitalist (VC) is an investor who invests in new and innovative companies with high potential to grow in exchange for a share of the company's ownership stake. Venture capitalists look for companies that are ready to commercialize their ideas. The terms venture capitalists and venture capital are used interchangeably. Jim Breyer, an early Facebook (FB) investor, and Peter Fenton, a Twitter investor, are two well-known venture capitalists. A venture capitalist (also known as a VC) is an individual or investment agency that invests in new enterprises or startups, and these venture capitalists are expected to offer management and technical knowledge, as well as a significant amount of money to their investments. A venture capital fund (commonly referred to as an LP or LLC[1]) is a collective investment entity that invests the financial resources of third-party investors in businesses that are too risky for traditional capital markets or bank loans.

---

[1] Limited Liability Company (LLC) & Limited Partner (LP)

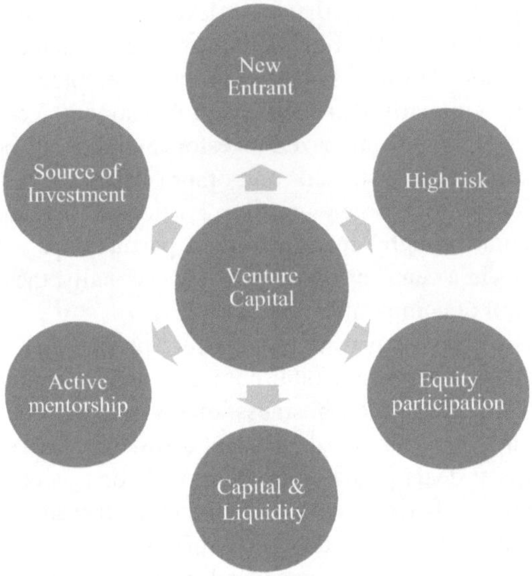

FIGURE 6.2 Features of venture capital. (Author's presentation.)

6.10 FEATURES OF VENTURE CAPITAL

As discussed earlier, the term "venture capital" refers to a type of private or institutional financing that can contribute a significant capital to growth-oriented businesses, mainly through equity acquisitions and occasionally through debt offers. Institutional investors who offer equity finance to nascent enterprises and actively advise their management are referred to as venture capitalists.

Furthermore, there are some features of venture capital or venture capitalist that also determine the working culture of venture capitalists. Figure 6.2 exhibits the features of venture capital.

6.10.1 New Entrant

Venture capital generally invests in new companies, which are established with the aim of high profits and spectacular returns. These companies introduce new products, processes, and management in the market and contribute to economic development and create avenues of employment.

6.10.2 High Risk Involved

As the VC invests capital in startups to kick start a business, which is not well established and profitable, it includes high risk and chances of failure. Although, before doing so, the VC evaluates the growth potential of a new company, still some risks are involved in it. As a result, it carries significant risks such as remanagement risks, product risks, market risks, and operational risks.

6.10.3 Equity Participation

As discussed in the previous section, the venture capital is a type of private equity participation. While investing in any new venture the venture capitalists share the equity participation in the investment. VC invests in the early stage of business and continuously provides funds to the entrepreneur to establish a business. A portion of equity participation gains and exit options are decided at the time of deal structuring. The exit of venture capital from the new venture is explained later.

6.10.4 Capital

The long-term capital is provided to a new venture to convert an entrepreneur's business plan or idea into physical form. So it helps the company to grow fast from the startup stage. A venture capitalist's primary goal is to achieve a higher profit on equity investments at the time of termination (exit), as well as a consistent return on debt funding.

6.10.5 Active Mentorship

Venture capital firms participate actively in delivering value-added services to investee companies, such as offering business skills and other services. They do not get involved in the daily activities of management of the companies they invest in, and do not own a majority or controlling stake in the companies they invest in. Moreover, they just act as mentors of new ventures by providing managerial support, monitoring, and follow-up assistance. They also keep a close eye on the physical and financial developments, as well as market expansion initiatives, which aids in the identification of essential resources and human capital to protect their (VC) investment in the company. The extension of hands-on management is justified by the fact that venture capital investments are notoriously non-liquid.

6.10.6 Mode of Investment

Venture capital works on the equity financing method, which is the amount of money invested in startups and small companies at a seed or early stage as they may not approach the capital market to raise funds. In addition, to assure a consistent return on the venture capitalists' investment, funding exits in the form of loan finance/convertible debt.

6.10.7 Liquidity

The achievement or failure of an innovative business enterprise or product determines the liquidity of venture capital investments. As a result, where new ventures are highly successful, there will be more liquidity. Investors seek ultimate capital gain when the investment is sold in the market. Enlistment of security assures the investments and its returns, or it is lost if the business is liquidated due to poor performance. It could take many years for the initial investment to start paying off. Sometimes, investments are locked for an indefinite period.

6.11 PARTICIPATION OF VENTURE CAPITAL IN AN ENTREPRENEURIAL COMPANY

Venture capital investment in an entrepreneurial company in India has four forms:

1. **Equity Participation:** Under this, ownership remains with the entrepreneur. Up to 49% of the equity capital is financed by venture capital.
2. **Conventional Loan:** Within this agreement, it charges a low fixed rate of interest until it begins commercial production. The loan must be repaid in accordance with the agreement once the standard rate of interest has been paid.
3. **Conditional Loan:** In such a loan no interest is paid. A conditional loan is repaid when the enterprise generates revenue in the form of royalties ranging from 2% to 15%.

6.12 FORMS/TYPES OF VENTURE CAPITAL

Venture capital is a funding to a new entrant in its business life cycle through various forms. Following Figure 6.3 shows various forms or types of venture capital:

6.13 VENTURE CAPITAL FUNDING PROCESS

There are various factors that venture capital investors consider such as, screening, contracting, monitoring, and selecting when to leave the investment (Wallmeroth et al., 2017). This section describes the venture capital funding process employed by the venture capitalists before funding any new venture. Figure 6.4 presents the funding process of venture capitalists. There are five phases in the venture capital process.

6.13.1 Deal Origination

A presentation of a business plan is the first step in approaching a venture capitalist. The following items should be included in the plan: (i) a summary of the business idea; (ii) future business opportunity, along with the size and potential of the market;

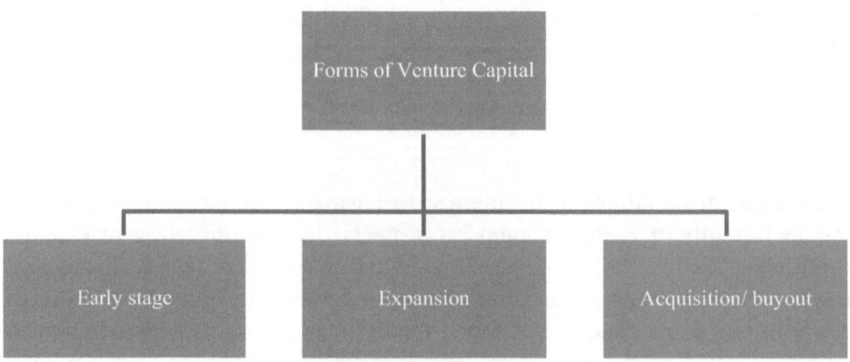

FIGURE 6.3 Forms/type of venture capital. (Author's presentation.)

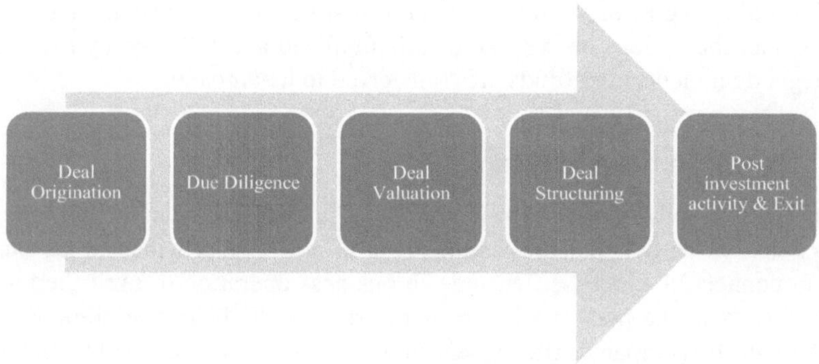

FIGURE 6.4 Venture capital funding process. (Author's presentation.)

(iii) examination of the current and anticipated competitive landscape; (iv) budget of the business plan in detail; and (v) description about the company's management. After receiving the proposal of a business plan by the entrepreneur, the venture capitalists do an initial screening of the business potential and decide whether to proceed with it or drop it.

6.13.2 Due Diligence

After receiving the business plan and performing the initial screening of the plan, if the proposal passes the initial screening, then the VC does a detailed evaluation or due diligence process. Here, VC examines the potentiality of the product, process, and technology of the proposed plan and also evaluates the personality traits and characteristics of entrepreneurs. VC investment is at high risk, so it assesses the risks and returns of the business plan.

6.13.3 Deal Valuation

The main motive behind the deal valuation is to perform an investment valuation process to agree on a satisfactory deal amount. The process includes the following steps:

- Forecasting future revenue and profit
- Expected enterprise value
- Determining the stake in equity structure based on the company's projected return on investment
- Considering various economic issues like demand/ supply of product, management team, unique selling point (USP's) of product, etc.

6.13.4 Deal Structuring

Once the venture capital is satisfied with the deal and its valuation, as well as finds the deal viable, the VC and the investee company negotiates the requisites of the deal that include funding amount and mode of investment. After getting ensured by the ownership stake, degree of VC involvement, investment risk, return, managerial performance, and the options of exit, the deal is finalized and followed by the completion of legal documents, and funds are transferred to a company.

6.13.5 Post Investment and Exit

A postinvestment activity means the degree of VC's involvement in the investee company. As cleared earlier that the venture capital not only provides funds but also assists in managerial expertise. During the business operation or until the business gets established in the market with great return, the VC helps the entrepreneur at every phase of the company's life cycle. Many challenges can be faced by the investment company such as low return on investment, less consumer attraction, technology damage, managerial function, and covering financial losses.

Exit refers to the way of exit of the VC from an investee company or a startup. The exit of the VC is possible in both cases whether the company is established with a projected profit or the company fails to achieve its motive. There are three modes of exit: IPO, buyback, and merger & acquisition.

- IPO
  When the company is well established in the market and performing very well with great profit, the VC decides to offer its shareholding to the public, and through this, there is a great chance that shares may value at a great premium.
- Buyback of Shares
  Under this exit, the investee company purchases the share of the venture capitalists. The advantage of this strategy is that the management control remains with the same company and the investee company gets more funds to expand its business. Thus, the company enjoys and maintains its reputation in the market.
- Merger & Acquisition
  In this strategy, another company that wants to expand its business in the same domain may approach the VC, and the VC can sell its holdings to that company. Through negotiating deals, the managerial control is transferred to the new company.

6.14 VENTURE CAPITAL FINANCING

Venture capital financing is a funding process in which the VC pools the money from the investors and invests the fund in new businesses that have the potential for high growth in the future. Venture capitalists invest in the new venture in the seed and later stage of a startup business. The following section describes the venture funding stages and the funding round of venture investment.

6.15 VENTURE CAPITAL FUNDING STAGES

Venture capital is a professional organization that pools money from other investors and professionally manages the fund. One of the major aspects of venture

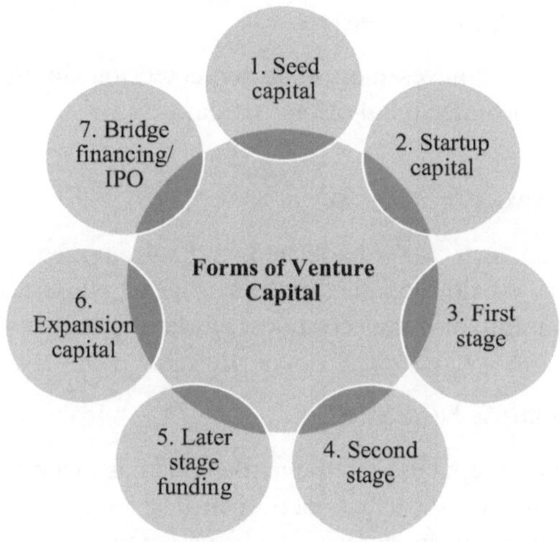

FIGURE 6.5 Stages in venture capital funding. (Author's presentation.)

capital financing is staged financing, under which companies are funded accordingly to the stage requirement. This type of investment is frequently used by VCs (Panda & Gopalaswamy, 2020). Venture capital funds businesses that are young, creative, and innovative. They help the new business promote its new products, procedures, and technological advancements. Figure 6.5 depicts the five stages of VC investment.

6.15.1 Stage 1. Seed Capital

This stage is the very first stage of financing. The VC invests or provides fund to the investment company which is in the very initial stage, not decided with sample product, process, market, and technology. Usually in this stage, the investment amount is very small. Here, VC helps the startup to develop its product, process, machinery, market research, and other administrative costs.

6.15.2 Stage 2. Startup Capital

At this stage once the business established with its sample product or it gets commercialized in the market, he entrepreneur needs capital for procurement of management, and for initial marketing to develop their products and services as per market demand. Investment at this stage is very rare.

6.15.3 Stage 3. First Stage

At this stage, companies start their operation and continue for at least 2–3 years. The business and management start functioning in the marketplace. At this level, VC investment may be able to assist you in increasing sales to the break-even point, enhancing productivity, or increasing the efficiency of a business.

6.15.4 Stage 4. Later Stage Funding

At this stage, company achieves more than projected and impressive revenue and sales, increasing the working capital of the company.

6.15.5 Stage 5. Expansion Capital

At this stage, the business is well established, and entrepreneurs search for a VC to get assistance to carry the development of the business to the next level. Here, funding might help expand into new areas or boost marketing efforts.

6.15.6 Stage 6. Bridge Financing

The company has a good reputation in the market and has made a lot of money. Now, venture capitalists look for ways to get out of their investee companies through IPOs, share buybacks, and mergers & acquisitions. If the company is planning an IPO, the VC will provide bridge financing to help with the costs of going public.

6.16 VENTURE CAPITAL FUNDING ROUND

6.16.1 Seed Funding

Seed funding mostly goes to product development and market research. As discussed in the earlier section, seed funding is a very early stage to get funds from VC. Here, the startup company gets the fund from various sources such as family & friends, crowdfunding, and investors.

6.16.2 Series Funding

Funding in series includes three stages: Series A, Series B, and Series C. There is not much difference between these series.

Under Series A, it takes place after a business is already operating in the market and people are buying and showing interest in a product. This means that the company is gaining market share and revenue is either steady or growing. At this funding stage, the investors do not look up to the business idea. They focus on how entrepreneurs plan to navigate the business in the market. Here, strategies are important to show how the business plan model will help gain more and more profit. The amount of funding in Series A is between $2 and $10 million.

In Series B and C, the companies are well established and the reason behind the need for funding is to increase market reach, larger customer reach, merge or acquire another company, etc.

6.17 VENTURE CAPITAL IN INDIA

The government of India recognizes the need for venture capital in the seventh 5-year plan and a long-term fiscal policy which was proposed in 1973 (Durrani, 2001).

FIGURE 6.6 Value of private equity and venture capital funding in India from 2015 to 2020 (in Billion US Dollars). (https://www.statista.com/statistics/880973/india-private-equity-and-venture-capital-investments/.)

The call for quicker venture capital as a source of funding/financing for new entrepreneurs, projects, and technology was highlighted by a committee on small- and mid-sized enterprise development. The Bhatt Committee, established in 1972 to identify the strategies for the development and improvement of small- and mid-sized enterprises, emphasized the need for venture capital in India, similar to the MacMillan Report in the United Kingdom in 1931. In acknowledgment to the obstacles faced by emerging entrepreneurs and technocrats in launching firms, the Committee proposed the creation of a Rs. 1000 million venture capital fund (Durrani, 2001; Chavda, 2014).

In India, venture capital financing began in 1988, with the establishment of Technology Development and Information Company of India Ltd., a joint venture between ICICI and UTI. Credit Capital Finance Corporation and Credit Capital Venture Fund sponsored the first venture to promote the Bank of India and Asian Development Bank, respectively (Chavda, 2014).

There was a continued increasing trend in venture capital investment in the fourth quarter, owing to consumer services sector-based firms – flipkart.com and snapdeal.com – which boosted their investment to 35% from the third quarter of 2014 (Shetty, 2017). The period from 1999 to 2016 experienced rapid growth and expansion of the venture capital industry. During this time, investment methods had also changed dramatically (Sabarinathan et al., 2017). There are many others sectors where venture capital increased such as healthcare, travel, real estate, and infrastructure. There was a 28% hike in venture capital funding in various sectors, in 2014. With burgeoning venture capital investment, 2015 also witnessed an emergence and evolution of VC funding in India. Compared to the year 2014 in 2015, there was 51% increase in the VC investment, and consumer technology was still a dominating sector for a venture capital funding (Shetty, 2017). Figure 6.6 depicts the growth of (private equity (P/E) & VC) funding in India.

Figure 6.6 presents the growth of venture capital and P/E & VC funding in India from 2015 to 2020. In 2020, there was an investment of 47.6 billion US dollars with 921 deals. Therefore, it presents that India is outperforming in the venture capital funding sector which experienced a rise of 28% in 2019 compared to 2018. In addition to this, in 2020, a total of $100 million was raised by 22 startups in India. Consumer technology, SaaS, and B2B commerce were the top three sectors that received approximately 75% VC investments in 2020 (Venture Capital Report, 2021). Figure 6.7 presents four phases of venture capital in India with major features of each phase.

6.18 VENTURE CAPITAL INVESTMENT IN INDIA

Many venture capital companies are running in India and funding many startups. Table 6.2 shows the top 10 active venture companies working in India.

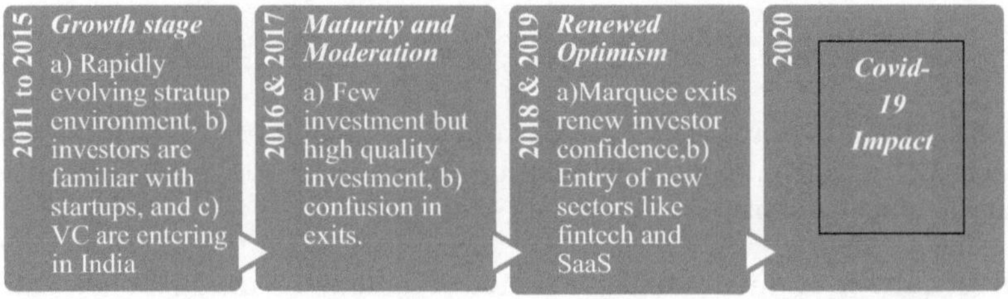

FIGURE 6.7 Four phases of venture capital in India. (Venture Capital Report 2021.)

TABLE 6.2
Venture Capital Investment in Companies in India

S. No.	Name of Venture Capitalist	Startup Funded
1.	Sequoia Capital India	Just dial, Knowlarity, etc.
2.	Helion Venture Partners	Make My Trip, Red Bus, YepMe, PubMatic, etc.
3.	Accel Partners	Flipkart, Book My Show, Myntra, Swiggy, etc.
4.	Nexus Venture Partners	Snapdeal, Komli, Housing, etc.
5.	Intel Capital India	Hungama, Snapdeal, etc.
6.	SAIF Partners	Paytm, Just dial, Homeshop 18, etc.
7.	Kalaari Capital	Myntra, Snapdeal, Zivame, Milkbasket etc.
8.	Tiger Global Management	Urban Company, Flipkart, OPEN, Razorpay, etc.
9.	Elevation Capital	Capital Float, Firstcry, Swiggy, Rivigo, etc.
10.	Blume Ventures	Dunzo, Unacademy, Procol, HealthAssure, etc.

Source: Author's own compilation.

6.19 CONCLUSION

For economic growth, entrepreneurial ventures work as 'engines' (Cassar, 2004). The chapter explains the venture capital financing methods and strategies in detail. To boost more and more startup in the country, venture capitalists play a very important role. VC finances the companies at every stage of the startup lifecycle. Venture capital has established itself as an essential financial market middleman, giving financing to new high-tech companies. Venture capitalists have their mechanism to provide funds to small-scale companies. The chapter also explains the evolution of venture capital evolution across India and the top ventures funded startups across India.

REFERENCES

Bellavitis, C., Filatotchev, I., Kamuriwo, D. S., & Vanacker, T. (2017). Entrepreneurial finance: New frontiers of research and practice. *Editorial for the Special Issue Embracing Entrepreneurial Funding Innovations*. 1–16. DOI: 10.1080/13691066.2016.1259733.

Bonini, S., & Capizzi, V. (2019). The role of venture capital in the emerging entrepreneurial finance ecosystem: Future threats and opportunities. *Venture Capital*, 21(2–3), 137–175.

Cassar, G. 2004. The financing of business start-ups. *Journal of Business Venturing* 19(2): 261–283.

Chavda, V. (2014). An overview on "venture capital financing" in India. *International Multidisciplinary Research Journal*, 1(2), 1–4.

Chemmanur, T. J., & Fulghieri, P. (2014). Entrepreneurial finance and innovation: An introduction and agenda for future research. *The Review of Financial Studies*, 27(1), 1–19.

Cumming, D., & Johan, S. (2017). The problems with and promise of entrepreneurial finance. *Strategic Entrepreneurship Journal*, 11(3), 357–370.

Denis, D. J. (2004). Entrepreneurial finance: An overview of the issues and evidence. *Journal of Corporate Finance*, 10(2), 301–326.

Durrani, M. (2001). *Venture Capital Financing in India: A Study of Venture Capitalist's Valuation, Structuring, and Monitoring Practices* (Doctoral dissertation, Loughborough University).

Panda, S. N., & Gopalaswamy, A. K. (2020). An analysis of timing decision in venture capital staged financing: Evidence from India. *Management Research Review*, 43(12).

Sabarinathan, G., Muralidhar, A., & Shetty, A. (2017). Venture capital and private equity investing in India: an exploratory study.

Sharma, R., & Rana, G. (2021). An overview of entrepreneurship development programs in India. *International Journal of Environmental Policy and Decision Making*, 3(1), 16. DOI: 10.1504/ijepdm.2021.119511.

Shetty, K. K. (2017). A comparative study on impact of venture capital financing on start-ups in India. In *International Conference on Education, Humanities and Management (ICEHM-17) March* (pp. 14–15). Statista report: https://www.statista.com/statistics/880973/india-private-equity-and-venture-capital-investments/.

Venture Capital Report, 2021. Retrieved from https://ivca.in/wp-content/uploads/2021/03/bain_report_india_venture_capital_2021.pdf.

Wallmeroth, Johannes and Wirtz, Peter and Groh, Alexander Peter(2017)., Venture Capital, Angel Financing, and Crowdfunding of Entrepreneurial Ventures: *A Literature Review Foundations and Trends in Entrepreneurship*: Vol. 14: No. 1, pp 1-129, 2018, DOI: 10.1561/0300000066.

Winton, A., & Yerramilli, V. (2008). Entrepreneurial finance: Banks versus venture capital. *Journal of Financial Economics*, 88(1), 51–79.

7 Woman Entrepreneurship in India

A Paradigm Shift

Ravindra Sharma
Swami Rama Himalayan University

Antima Sharma
University of Engineering & Management

Geeta Rana
Swami Rama Himalayan University

Vipul Jain
Shri Guru Ram Rai University

CONTENTS

DOI: 10.1201/9781003217084-7

7.1 INTRODUCTION

In the early 1990s, research on entrepreneurship was not segregated as male or female, it was, by default, entrepreneurship meant male entrepreneur. Research considered that entrepreneurship is generally the same for both males and females, and there is no need for separate study (Bruni et al., 2004). Therefore, women entrepreneurship was not recognized and developed until the late 1990s, and the subject of woman entrepreneurship was uplifted when the conference was held by Organization for Economic Cooperation and Development (OECD) in 1998 (Yadav & Unni, 2016). In the last few decades, woman entrepreneurship has shown immense growth in both the market and in the literature. Women showed great participation in starting business and running it successfully, woman-owned business play a prominent role in developing the society, economic condition of a nation, and help in generating employment opportunities. It not only contributes economically but is also recognized as a source of increasing job creation, growth, and women empowerment (Audretsch, Keilbach & Lehmann, 2006). According to literature, women play an important part in the broader entrepreneurial and economic growth phenomena (Sarfaraz et al., 2014; Sharma & Rana 2021a).

Woman entrepreneurial is now visible in the business environment, this is because they have played a vibrant role as policymakers, practitioners, and researchers all over the world. In addition, the changing business landscape associated with dynamic changes leads to high levels of unpredictability of the business environment. Women and men perform differently in terms of entrepreneurship, according to empirical findings. In terms of innovativeness, proactiveness, and risk-taking, it was discovered that gender had a substantial impact on entrepreneurial activity in businesses. Many scholars worked to investigate the different dimensions of women's entrepreneurship and provide a better understanding of its concept.

> The empowerment of women is powerful beyond measure and beautiful beyond description.
>
> *Steve Maraboli*

Entrepreneurship of women is the way through which women manage all the factors of production and are ready to take risks as well as involve in creation and innovation activities and generate employment for others (Sharma, 2013).

In general, woman entrepreneurship is a process in which women take the initiative to establish a business risk-taking, acquiring resources, confronting obstacles, and employing people. Women's entrepreneurship may be described as a woman or a group of women who start, organize, and run a company. To be labeled a woman entrepreneur, women must invent, imitate, or adopt an economic activity. In technical terms, women's entrepreneurship is defined as an enterprise owned managed by a woman who has a minimum financial capability of at least 51% of the capital.

According to the OECD (2015) report, women's empowerment is a precondition for long-term development and pro-poor growth, particularly in emerging economies. As 50% of the world's population is constituted by women, there is a need to consider the economic empowerment of the woman for the nation's economic upliftment.

Many countries work for the upliftment of women as an entrepreneur to develop the economic condition of the nation, and India is one of the nations focusing toward this direction and making efforts to increase the number of woman entrepreneurs.

7.2 OBJECTIVE OF THE STUDY

1. To understand the idea of women entrepreneurship in the Indian context and to investigate the variables that encourage women to start businesses in India.
2. To identify the problems and barriers that women face as an entrepreneur in the Indian context.
3. To discuss a successful story of an Indian woman entrepreneur.

7.3 RESEARCH METHODOLOGY

This study is qualitative, and the analyses and conclusion are based on the data collected from various research papers, books, journals, websites, published and unpublished works, articles, government policies, and work performed in this field.

7.3.1 Research and Purpose

Considerable research was conducted to study about entrepreneurship and its activities. Many social and economic entrepreneurial activities are recognized by various governments around the world are trying to foster entrepreneurship (Dana & Dana, 2005; Sharma & Rana, 2021b). Studies conducted focused on women entrepreneurship, yet there is enough scope for its exploration, especially in India (Kaushik, 2013; Korreck, 2019).The qualitative method used by the researcher is summarized on the basis of personal observation of situations, events, individuals, interactions, and transactions, as well as document analysis (including quantitative records), and open-ended interviews yielding in-depth and oral testimonies.

7.4 LITERATURE REVIEW

In recent years, an extensive literature about entrepreneurship has been published; hence, this subject is not new. Several studies were conducted to define women entrepreneurship (Kumar and Gupta, 2013), which defined a woman entrepreneur as a person or group of women who performs a challenging role by running a business to meet their daily needs and financial independence. Extensive literature on women entrepreneurship both domestic and international is available, yet there is much more to explore in this area. Some of the main studies have been mentioned as follows

Garg and Agarwal (2017): the researchers found that Indian women are working in the field of manufacturing, trading, and service sector from the early situation where they were limited to only domestic jobs. Women face major issues from society followed by finance, lack of resources and knowledge, and the work environment. The problem can be solved when a good work environment is created by society and the family for women to start entrepreneurship (Idris and Tan, 2017). In this study, the researcher focused on different motivation for women to start the business. The study concludes that women want to start their own business because they want greater freedom, financial independence, flexibility in their life, self-satisfaction, respect and identity in society, and show their creativity. The study by Khaire (2011) confirmed that entrepreneurial venture by women leads to the development of entrepreneurship and other aspects of business in women. The study also observed that work has been conducted from a policy implementation point of view only in aspects such as psychology and behavior of women are out of scope.

Ojong et al. (2021) analyzed articles and found how cultural, institutional, economic, political, and social contexts influence the resources and strategies used by female entrepreneurs, and how interactions between contexts, resources, and strategies determine the outcomes of female entrepreneurship in Africa. We suggest that the circumstances in which female entrepreneurs thrive contribute to the development of new technologies. Anggadwita (2021) analyzed the influence of the sociocultural environment, social perceptions, and entrepreneurial orientation on women's entrepreneurial intentions. The purpose of this study is to show how female involvement in social entrepreneurship may show more influence on the creation of social values and show eastern tradition (Anggahegari et al., 2021). The study focuses on identifying entrepreneurial intention in women as an entrepreneur who owns small and medium-sized enterprises (SMEs) in Indonesia. The study results confirm that sociocultural and social perceptions have a significant effect on women's entrepreneurial intention. The chapter opens with an overview of North Macedonia before moving on to an evidence-based assessment of women's entrepreneurship in the country. The study proceeds to discuss government and other policies, initiatives, and activities aimed at encouraging women to start businesses. Finally, a forecast for the future growth of women's entrepreneurship is provided (Ramesh, 2018). Indian government is realizing the importance of women entrepreneurs in the economies of country. The government is trying to build a culture for women entrepreneurship, yet it has failed to develop this because of the lack of a proper environment. On the basis of the available studies, Saraswat and Lathabahvan (2020) concluded that discussion and personal opinions would help them to consider entrepreneurship for their living and create job opportunities for others. Case studies will stir their minds and lead them to unimaginable heights. The whole motive of this study is to say that women can contribute to the economy. Tiwari's (2017) study had put light on the diversified profile of women entrepreneurship in India. The study found that numerous challenges are faced by women entrepreneurs in the course of their careers and there is a need for comprehensive action plans to resolve problems faced by women as an entrepreneur. Women entrepreneurship is a diverse and complex domain, which requires extensive and intensive research endeavors for decoding its dynamics. The study by Yadav and Unni (2016) examined a number of studies published on women's entrepreneurship and found that there is a rapid growth in the studies of women entrepreneurship. According to this finding, there is still a long way to go in terms of developing a robust theoretical foundation for research on women's entrepreneurship.

7.5 WOMAN ENTREPRENEURSHIP IN INDIA

There is a shift in society with women entering the entrepreneurship world. In early Indian culture, women were mainly concerned with home affairs and doing some job as per the convenience of the family. After the 19th century, India has been observing a change in women's work profile from the kitchen to professional fields. In the present scenario, women are showing their presence in every field such as trading, industry, engineering, and so on. It is a very much remarkable sign that women are contributing to the nation's growth. After globalization, changes in the country's culture and society were observed and the mentality has changed about women and their roles have also been modified. A few studies statistically confirmed that women's participation in small businesses has increased in recent years, even in wealthy nations such as the United States and Canada. Now a woman is not to be kept at home (Kaviarasu et al., 2018; Shivani, 2021). The enhancement in women's position and power is due to advancement in education, awareness, and support provided by family, society, and most importantly, the government.

7.5.1 Facilitators to Promote Woman Entrepreneurship in India

In couple a of decades, many studies will draw the attention of government and non-government bodies on significant factors, such as education, resilience, personal satisfaction, and providing employment, which decide the success and failure of women entrepreneurs (Agarwal, 2021). Studies conducted and drawn attention to overcome the challenges faced in the development of women entrepreneurs. According to the results of the study, it is found that the main factor that encourages women to be entrepreneur is the freedom to take decisions. Women's Entrepreneurship Development Process in India is rapidly being recognized as a major untapped source of economic growth, and therefore, women entrepreneurs generate new job opportunities and pathways to economic independence for women (Devi & Parimala 2013). Although the Indian society has always been a male-dominated society, women have crossed hurdles and utilized opportunities and become most memorable and inspirational entrepreneurs (Kaviarasu et al., 2018). However, in the sixth economic census issued by the Ministry of Statistics and Program Implementation, women account for around 14% of India's overall entrepreneur base or 8.05 million out of a total of 58.5 million. The average number of employees employed by a woman-owned firm is only 1.67. In India, the proportion of female entrepreneurs is increasing. The share of female entrepreneurs in India is showing a healthy growth of 71% per year (Chaudhary, 2017). Nowadays, woman's potential as an entrepreneur is increasing in the Indian society because some factors are favoring women to become a successful entrepreneurs, and after reviewing of the literature, some factors are as follows:

Government Schemes and Programs: The government and government schemes play a major role in the development of women entrepreneurs in India. Through various schemes, programs, and education, creating awareness for women's upliftment and promoting them for self-employment as per their convenience. Here are some of the schemes provided by the government to help women entrepreneur financially to start their own businesses:

TABLE 7.1
Schemes by Government for Woman Entrepreneur

Bharatiya Mahila Bank	Bharat Mahila bank started as a bank that dedicated to helping disadvantaged women develop companies' worth up to 20 crores. In 2017, this plan was put in place and eventually joined with SBI.
Annapurana Schemes	This program is for female entrepreneurs who have opened a food catering business. They can get up to Rs. 50,000 in loans to buy cooking equipment and water filters. The loan can be repaid in 36 installments.
Stree Shakthi	This scheme applies to women entrepreneurs who have majority ownership over 50% in a small business. This scheme offers an interest concession of 0.05% can be availed on loans above 2 lakhs.
Orient Mahila Vikas Yojana Scheme	This scheme is available for women entrepreneurs to facilitate a loan of up to Rs. 25 lakhs. The repayment of the loan within 7 years. The rate of interest in the loan is 2%.
Dena Shakyhi Scheme	This scheme offers loan for women entrepreneurs for those who want to do a business in manufacturing goods and agricultural. Loan available of Rs. 50,000–20 lakhs and offered concession of 0.25%.
TREAP (trade-related entrepreneurship assistance and development scheme)	This scheme to empower women by giving credit for initiatives, offering specialized training and counseling, and gathering data on relevant needs. This program allows financing institutions to appraise up to 30% of the overall project cost. The remaining 70% would be financed by the institutions.

Source: S. Saranya & Dr. K. Chandrashekhar (2017).

Table 7.1 presents some schemes run by the government to provide financial assistance or to support those women who want to start their own businesses. The government also provides concession and rebate to repay the loan and in the interest rate. The government is working for the empowerment of women from the very beginning with a 5-year planning, **First Five Year** (1951–1956)—Welfare-Oriented Program; **Second Five Year** (1956–1961)—woman welfare programs such as equal pay for equal work; **Third Five Year** (1961–1966)—education as strategy for woman welfare; **Fourth Five Year** (1969–1974)—to promote family welfare within the family and education; **Fifth Five Year** (1974–1979)—to develop the woman in society; **Sixth Five Year** (1980–1985)—increasing enrollment of the girls at the elementary level for development and equality of woman; **Seventh Five Year** (1985–1990)—Women Mahila Mandals were established. Many arts and craft centers were opened for women to enhance their employment opportunities; **Eighth Five Year** (1992–1997)—development to empowerment; **Ninth Five Year** (1997–2002)—vocational training programs for woman; **Tenth Five Year** (2002–2007)—promotion for self-help group programs of woman; **Eleventh Five Year** (2007–2012)—providing environment like violence against women, economic empowerment and women's health so women can work in Indian society. **Twelfth Five Year** (2012–2017)—women participation in related work, both in agriculture and animal husbandry and pushing single woman (Kaur, 2018). This depicts that the government of India has been working since 1950s to make strong base for the woman to raise them in society equally, and this is the main factor that facilitates most woman to work as an entrepreneur.

MSME: The Ministry of Medium and Small Enterprises has continuously been working for woman empowerment since the 1950s, which carries various clauses. The ambiguity over the status of a woman was eliminated when the Oxford dictionary declared "Nari Shakti" as a Hindi word for 2018. The Ministry of MSME is attempting to empower women entrepreneurs through several programs that encourage women's potential and assist them to develop their identities. Since its commencement, 138,000 enterprises have been started by women entrepreneurs under the Prime Minister's Employment Generation Program (PMEGP) scheme. Women entrepreneurs are responsible for approximately 30% of the overall initiatives funded by the PMEGP. Under the strategy, women entrepreneurs fall into a special category and are eligible for 25% and 35% subsidies for projects in urban and rural areas, respectively. Own contribution for women recipients is only 5% of the project cost, whereas it is 10% for the general group. During the 2016–2017 and 2017–2018 academic years (msme.gov.in), these programs help to motivate women to come in front and show their skills and capability. There are some major highlights made by medium and small size enterprises (MSME) to promote woman entrepreneurship in the year 2019–2020.

Woman Herself: The woman herself is the biggest factor to become an entrepreneur, she convinced the society of her potential to accept them as an equal partner in the economic development of the country. Today they are performing outstandingly in all the fields such as a designer, decorators, publishers, and garment manufacturers, and trying to touch new avenues to get equality in society. Here are some of the characteristics that woman should have to be a successful entrepreneur.

Education and Awareness: Apart from formal motivational factors provided by the government bodies, education is the most prominent reason to encourage women to become self-employed and to come out of their old image of being a homemaker only. Education and knowledge give sense to judge what is right and wrong, an educated woman will be able to make the right decision at the right time. In India, Kerala has the highest literacy rate (including women's literacy), indicating a favorable environment for the formation and growth of women entrepreneurs in the state (Sadhin, 2019).

Institutional Support: With the government and MSME, many institutes, non-government organizations (NGOs), banks, and other bodies also putting efforts for the development of woman entrepreneurs. Banks and financial institutes provide flexible financial schemes for the women entrepreneurs, they trying to achieve that every girl should have bank account so they can access loan facility. NGOs supporting women through boosting their motivation and giving them access of market for their product such as haat, seminars, workshops, and so on. NGOs play a significant role to execute

Some of the major promotional schemes for women entrepreneurship are shown in Figure 7.1.

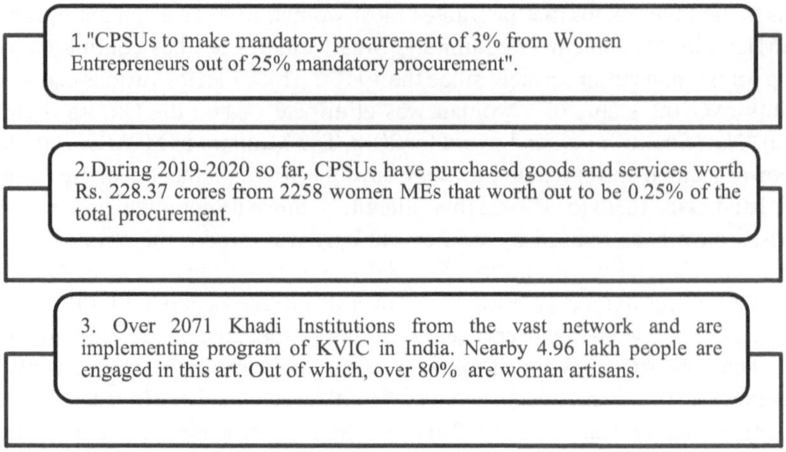

FIGURE 7.1 Major promotional schemes for women entrepreneurship. 1. Box = Describes step taken by (*CPSU) Central Public Sector Unit for Women Entrepreneurship (3% of Mandatory Procurement from Women Entrepreneurism); 2. Box = (*CPSU) Central Public Sector Unit for Women Entrepreneurship procured worth 0.25% good from women MSEs; 3. Box = To promote artisans government introduced program (*KVIC) Khadi and Village Industries Commission in India. In that more than 80% are woman artisans.

FIGURE 7.2 Characteristics of women that supports the women to become entrepreneurs. (Self and Inferences from Goyal and Prakash, 2011.)

The above figure 7.2 is showing the characteristics which are vital for women to become successful entrepreneurs.

government five-year plan policies. Universities and Colleges help woman to provide them vocational training and give awareness about technological advancement. Some of the known institution working to magnify the role of woman in entrepreneurship is:

1. **Federation of Indian Women Entrepreneurs (FIWE):** It was founded in 1993 after a resolution was passed at the fourth International Conference of women entrepreneurs (WE) at Hyderabad. FIWE associates with many other associations to facilitate members in their work.
2. **Small Industries Development Bank of India (SIDBI):** It is an institution that facilitates small-scale industries. SIDBI has introduced specific schemes for women.
3. **Khadi and Village Industries Commission (KVIC):** It is established under the Khadi and Village Industries Commission Act, 1956, and it works to provide employment opportunities in rural areas and for women entrepreneurs.
4. **National Bank for Agriculture and Rural Development (NABARD):** Liberal credit policies given to rural women entrepreneurs.
5. **International Centre for Entrepreneurship and Career Development (ICECD):** It works internationally and provides training to woman entrepreneurs to support MSME over the last 25 years in above 55 countries.
6. **National Alliance of Young Entrepreneurs:** It was set up in 1975 for the promotion of women entrepreneurship. It is the most representative organization of women entrepreneurship in India.
7. **Mahila Kalyan Smiti:** It helps to provide financial grants to women through government schemes and also with the help of the Ministry of Textile, they provide training on handicrafts, jute, and handloom products. She provides rural women education and jobs, and she encourages them to attend workshops, seminars, and exhibits to learn about business formation, entrepreneurship, and market access (Lenka & Agarwal, 2017).

Society Support: Along with the government's schemes, programs, policies, and other institutional society is also accepting the role of women in economic development. Family, friends, relatives, offices, industries, associations, and nearby people are understanding the role of women in working and accepting it. Today, the cost of living is very high and only men's earning is not enough to survive; there is a need for contribution of both men and women. Therefore, families started encouraging working of women, and it was noticed that women were more comfortable with entrepreneurship compared to 9 AM to 5 PM jobs.

7.6 BARRIERS FACED BY WOMEN TO BECOME AN ENTREPRENEUR IN INDIA

After so many supporting factors, still women entrepreneurship is not still flourishing in India at its optimum level and even though entrepreneurship has its own advantages. It is due to some reasons and problems, which are faced by women. There are two main factors considered as challenges for women entrepreneurs: economic and social (Samiti, 2006). A major obstacle faced by women is maintaining balance between family obligations and work commitments. Numerous problems faced by women entrepreneurs such as finances, lack of time, fixing priority between family and work, and lack of participation in training program. Many studies have reported some of the barriers that affect the success of women entrepreneurs such as:

Balanced between Family and Career: Women are overburdened with family responsibilities, and it sometimes is difficult to put complete efforts in her business affairs.

Financial Constraints: For all the entrepreneurs to start their own setup, finance is their main difficulty for them. Although the government is affiliating it through various credit schemes, woman are getting fewer opportunities than men to get access because of the lack of collateral, not accepting household assets easily, and lack of trust in women in the society (Mahbub, 2000).

Lack of Education and Awareness: Very few educated Indian women are running their own businesses; generally, educated women prefer regular jobs. Most of the start-ups are run by uneducated and rural women, therefore, because of illiteracy, they face other problems and are not aware of benefits. Most women entrepreneurs are unaware of the advancement of new technology, new techniques of manufacturing, marketing, networking, and other governmental backing that will enable them to advance in the field of management due to a lack of adequate education (Rudhumbu et al., 2020).

Lack of Exploration: Women start their business and do it at their convenience, they do not want to explore it more because of the above discussed reasons. They restrict and limit themselves to some limited activities; therefore, they face issues such as lack of confidence, not obtaining proper entrepreneurial skills, lack of interest in making new business networks, and sometimes failed to break their image in society too (Rana & Sharma, 2017).

Unfavorable Environment: The situation and condition sometimes became unfavorable for the women at work, due to this, women are unable to run their business successfully (Agarwal, 2020).

7.7 IS COVID A CHALLENGE OR AN OPPORTUNITY FOR WOMEN ENTREPRENEURS?

COVID-19 has disturbed mankind and all sectors all over the world. Many people have become jobless and not able to obtain another, women also faced the same. On other hand, COVID-19 has given opportunities to jobless people to discover the solution to serve the society and fulfill their needs in this pandemic with their experience, innovative ideas, and skills. During a crisis, people's ability to stay connected is critical. Businesses from all across the world have banded together to assist their communities in unprecedented ways. Despite physical barriers, entrepreneurs have found solutions to the problems that have arisen as a result of the epidemic. Even in the face of enormous sorrow, business leaders, researchers, and scientists have found a way to promote connection (Milan, 2020).

Women have shown their venerable contribution and interest in this pandemic by starting small economic activity through their talent, innovative thinking, limited resources, experiences, and with the help of family members. While research on women entrepreneurs who pivot their company models to take advantage of the opportunity provided by the COVID-19 pandemic is still in its early stages, the popular press is filled with examples of bright women who are responding to COVID-19's problems in innovative and distinctive ways (Manolova et al., 2020). If we observe our nearby also, woman those who lost their jobs and started small business with their experiences, knowledge, creative ideas, and interest such as online teaching classes, cookery shows, mask making, garments designing, tiffin center, homemade papad and pickles, and so on. After going through research on impact of COVID-19 on entrepreneurship, it is found that woman entrepreneur is not get affected much by this pandemic because:

1. Most of the women entrepreneurs doing business, which caters day-to-day necessity of the society.
2. Most women start businesses that do not require much investment and they use their own savings rather than depend on loans and other financial assistance.
3. Women are able to manage with limited resources.
4. Most women carry forward their own interest and hobby as business, so they do not give up easily.

So this COVID-19 pandemic did not bothered woman entrepreneurs much and gave opportunities new or existing one to take advantage of this opportunity:

> Pooja Dhingra was running a successful café by the name Lecl5cafe & Patisserie, Mumbaiv, but due to the lockdown she has to shut down that, lose everything for which she has worked over the decades. Pooja has not given up and, she started baking at home. Pooja Dhingra mentioned that:
>
> "We got our heads down and released an e-book to meet the salary of our staff, and she created a variety of boxed cake mixes and "dipkies" – biscuits to be dipped in hot beverages." I started taking online workshops as well," Dhingra added.
>
> "This doesn't happen overnight," she added. "In the past, I would work on something until it fit my image of perfection. With the pandemic, I understand that I can't take another day for granted. If something needs to be done, it should be done today."

Another story of Rajasthani community, banjara's having a famous flea market and selling antiques and vintage furniture witnessed the immediate negative impact of the lockdown. Srishti Tehri, a 25-year-old fashion media ghost once passed through marked and noticed that there was no single customer who visited that market in lockdown, she was the only customer of the day.

> Tehri felt obliged to act almost immediately. She came home with photographs of the market's ceramics, created a logo, and shared the photos on Instagram in the hopes of attracting attention.

The response was tremendous. Tehri received 131 orders and sold over 800 things in her first sale. The "Artisans of Banjara" was formally launched by Tehri.

Most entrepreneurs' businesses have faced a decline and are yet to return to pre-COVID-19 levels. Only selected entrepreneurs are able to take a stand again previous position, despite losses and risks, entrepreneur showed greater optimism. In addition, among businesses, women have proven to be adaptable, with 54% having already adjusted their company model in reaction to COVID-19, and another 24% planning to do so by December, according to an interview. Only 6% of respondents believe that they will be unable to change their business model. Women's sole obstacle to changing the company model was a lack of expertise about new products, markets, and management techniques.

COVID-19 has given an opportunity to work from home or operate from home, and this has shown mixed impact on women entrepreneurs. As per a study, 57% woman experienced an increase in productivity while working from home and 43% noticed negative impact on productivity due to working from home.

(Statistical source: Report by Megha Chawla, Prapti Sahani, Kamyani Sadhwani interviewed 350 women solopreneurs of the Urban area to know the impact of COVID 19 and challenges face by them.)

As per the report in Economic Times (October 14, 2020), 90% of total women entrepreneurs and especially in urban area are confident that they will recover soon from their losses and come back to their previous position.

All associations working for women trade such as the Federation of Indian Women Entrepreneurs, All India Hair and Beauty Association, Consortium of Women Entrepreneurs of India and ITF-Industry and Trade Forum have conducted a survey on the impact of COVID-19 and suggested government the same. The report mentioned that most of the women entrepreneurs are in the informal sector and they are not able to avail government schemes and obtain assistance.

The government should work on this area and provide them with loans, concessional rates, provide mentors and awareness among them and also provide transportation facility so that they can reach a bigger market to sell their products (Report of The Hindu, August 27, 2020).

7.8 SUCCESSFUL WOMEN ENTREPRENEURS IN INDIA

One should be proud to be a woman. Women's success tales abound in every sector of life, be it corporate management, household management, or the sporting world. A majority of women are born with business abilities and are well known to them. Patience, perseverance, risk-taking ability, hard work, and other qualities are ingrained in women (Jedy, 2017).

Indian women have shown remarkable progress as an entrepreneur not nationally but internationally too. Women who are not from urban area but from rural area, less educated, and under privileged have also emerged as successful woman entrepreneurs. Women have a presence in almost all sectors commerce/ecommerce, education fashion, aviation, kitchen, retail, investing, they proceed with courage and enthusiasm to make a difference in society.

The Prime Minister of India also encouraged women in trade activities and contribution to Indian economics; he cited many examples of women entrepreneurs who have made an impact on both state and national levels:

Jassuben Pizza

Jassuben is basically from Gujrat and enjoys being a homemaker, and she used to make pizza for her children. This idea was a success when people started appreciating her pizza. People started preferring "Jasuben Pizza over Pizza Hut." After Gujarat's Chief Minister, Narendra Modi, named it a good example of female entrepreneurship in his speech at FICCI in 2013, it received a lot of media attention.

Lijjat Papd

Another example mentioned by Mr. Narendra Modi is Lizzat Papad. It was started by seven women who borrowed rupee 80 and initially made four packets of papads. Shri Mahila Griha Udyog manufactures many products from papad to appalam, masala, gehu atta, chapati, SASA Detergent Powder, SASA detergent cake (Tikia), and SASA liquid detergent. Today, the co-operative has annual sales of more than 3.1 billion.

COWE

Mr. Narendra Modi also addressed the Confederation of Women Entrepreneurs (COWE) as a perfect example of women empowerment; it is an association that works for the upliftment of women through workshops, seminars, and counseling. Rupa Rani, founder of COWE, fulfilled the dream of many women.

All these stories also inspire many women of the country who want to start their own business. Their struggles give strength to others to face the challenges with enthusiasm; there are so many other successful stories of today's women entrepreneurs or 21st-century women entrepreneurs, and some of them are as follows:

Falguni Nayyar: First, she was a managing director in Kotak Mahindra Group and after completing a successful career of 25 years, she left the bank. After that, she started a business by named NYKAA.COM, which is a premium beauty and wellness platform for women nowadays. She is an MBA graduate from IIM Ahmadabad, and with her knowledge and skills, she made NYKAA as a top brand in beauty and cosmetics.

Meghna Agarwal: She is a co-founder of Indi Qube, doing well with the idea of subleasing new entrepreneurs by providing vacant space with facilities such as gym, food transport, and so on. Indi Qube enjoys a good market for its services, and it received $ 15 million in 2018 through West Bridge capital.

Sairee Chahal: She is the founder of SHEROES and the co-founder of Fleximoms and Newslink. She is known as a serial business woman in India. She introduced the platform for woman to facilitate them with resources, opportunities, and interactions. Sairee was awarded the Most Powerful Women in Indian Business.

"Core philosophy of theirs is to help women navigate career stages and find work that fits their life".

There is a long list of women in India who are showing their entrepreneurship skills in almost all sectors (see Table 7.2). As we have discussed the stories of successful

TABLE 7.2
Statewise List of Women Entrepreneurs

Entrepreneurs	Creations	State	Mission
Dinika Bhatia	DRB Foods Pvt Ltd.	Delhi	Created Benchmark with Premium Dry Foods-Consumer Packaged Goods
Dipika Ajay Khitan	Maiya Publishing	Bangalore, Karnataka	Boutique Publishing House-Services for authors to publish a dream book
Hemalatha Annamalai	Ampere Vehicles	Coimbatore, Tamil Nadu	The first company in India manufactures key components of electrical vehicle
Divya Seth	Wearable Arts	Kolkata, West Bengal	The producer of wearable art and customers can invest into beauty product
Garima Agarwal	MN Design Co	Lucknow, Uttar Pradesh	Create affordable customized luxurious for her clients "You dream we create"
Geetnajali Kasliwal	Anantaya	Jaipur, Rajasthan	Handmade luxurious item
Gorsi A. Parekh	Arraycom	Gandhinagar, Gujarat	Telecom integrated and seismology industries
Meeta Mehra	Divine Chocolate Factory	Amritsar, Punjab	Home Kitchen Turned Café
Mannjori	Midori, Connecting Through Craft	Guwahati, Assam	Global Business Platform for North Eastern crafts
Nazneen Katrak	Mustang Socks	Palghar, Maharashtra	Domestic made socks
Payalh Agarwal	Zelena Tanquilitea	Gurgaon, Haryana	Tea portfolio
Sunita Mohanty	SMA e-Experts Private	Bhubaneshwar, Odisha	Leading risk management company

TABLE 7.3
Findings of the Study

S. No.	Objectives	Findings of the Study
1.	To know the concept of women entrepreneurship in the Indian context and to explore about factors that support woman entrepreneurship to flourish in India	Slowly and steadily, the contribution of women is increasing in the development of the Indian economy. Main facilitators in to promote women entrepreneurship in India are education, society development and awareness programs run by government bodies, institute and NGOs
2.	To identify the problems and challenges faced by woman as an entrepreneur in Indian condition	Apart from common issues faced by an entrepreneur, a woman also faces the issues like family foundations, limited mobility, less exploration, lack of trust of investors and stakeholders, and low-risk-bearing capacity, lack of promotional programs and schemes, and so on
3.	To discuss the successful story of Indian woman entrepreneur	Kalpana Saroj, Kiran Mazumdar, Indra Nooyi, Falguni Nayar, Ritu Kumar, Jassu Ben, and so on are some of women who set an inspiring example for other women

women in the field of entrepreneurship, it shows that, these days, women are equipping themselves with a large contribution in overall economic of a nation. The largest contribution from women entrepreneurship is from the southern states of India. Most of the women own enterprises in the education and manufacturing sectors (GIZ).

7.9 FINDINGS OF THE STUDY

After reviewing and analyzing the extensive literature and research work performed in the field of women entrepreneurship in a special context to the Indian scenario, it is found that:

Table 7.3 shows that findings of the study that covers the objective of the study. In India, there is need to make more efforts to bring those women as an entrepreneur in the main frame. More facilitating programs to run, so women can easily make up their mind to start their own small enterprises as per their convenience.

7.10 DISCUSSION

Worldwide, more than 163 million woman entrepreneurs are changing rapidly the start-up communities. With the speed women are capturing the business world, the word CEO will soon be a nomenclature "SHE-E-O" (ET Goverments.com 2020), India is also not aloof from this creative change, there are a few names like Lijjat Papad and Amul Cooperatives of women entrepreneurship. Nowadays, Indian women are not limited to only small and medium sectors, but they entered in the others main stream sectors too such as artificial intelligence (Rana and Sharma, 2019), space sciences, robotics, fashion world, food industry, electronics, medical science, and service sector.

According to interview partners, women entrepreneurs generally work in social sectors and industries including health, clean energy, and zero waste, education, women's hygiene, fashion, cosmetics, food and nutrition, garments and textiles, and management and human resources (Ellina and Tomar, 2018). The top five states with the highest percentage of women entrepreneurs are Tamil Nadu, Kerala, Andhra Pradesh, West Bengal, and Maharashtra (Korreck, 2019). According to research done in Kerala and Tamil Nadu, becoming a woman entrepreneur was seen positively by 28% of respondents in Kerala (Rana and Sharma, 2019).

From the review of different studies on woman entrepreneurs, it is found that women became entrepreneurs to find their freedom, to become self-dependent, due to domestic violence or disrespect, awareness and high education of women and programs in support of woman education like "Beti Padhao Beti Bachao," workshop, and seminar organized by NGOs and personal interest also. The different studies suggest that one factors that facilitate woman entrepreneurship is woman herself. Entrepreneurial women who are capable of providing values in both home and social life have a strong desire to accomplish something positive ingrained into their DNA. Women are more aware of their own characteristics, rights, and job conditions as a result of the media (Sivanesan, 2014). The study found that the availability of leisure time motivated women entrepreneurs from higher income classes (Rani, 1996). Other supporting factors are government/MSME schemes, NGO/institutions role, awareness programs, changed behavior of society, and so on.

"The days of women following the crowd or being afraid to go further than them are long gone. Today's women are unstoppable, having made significant contributions to every field from culinary arts to theater, marketing to science and technology" (Slama Siddiqi). They have long battled all the problems of women entrepreneurs to be where they are today, but still facing some challenges to start their own business mainly to balance between family and business, less trust of investors on woman entrepreneur, still not proper education system for woman, not fully developed mindset of society about women's growth, less networking and tradition and culture of Indian society to not give credit too woman, and so on.

> The entrepreneurial journey is full of challenges in itself. But women entrepreneurs not just have to face the problems of entrepreneurship but also of being a women entrepreneur. However, many women nip their interests in the bud or give up their dream of becoming a business woman midway.
>
> *myloancare.in (2019)*

After all challenges and problems, many Indian women has made a mark in the business world and also inspiring others in society. "Today, we are in a better situation, since women's engagement in the sphere of entrepreneurship is rising at a significant rate. Nearly 45% of India's population is made up of women." (Singh, 2014). Women entrepreneurship have been made an appreciable impact on the economy of nation, that is more than 25% of all kinds of business. A review of the studies confirmed that

TABLE 7.4
Women Who Are Empowering SME IN India Through Their Services

Women Entrepreneurs	Head of Organization	Services Provided
Shristi Banka	Banka & Banka	Advisory firm—Strategic support to MSMEs and start-up eco system
Arundhanti Mukherjee	Aaroh	Marketing Consultant Firm-"NewNormal" provides full range of services to establish a start-up in market, work on market capture
Pitha Datta Chowdhury	Econolytics	Data analytics—data mining for business purpose in the affordable price
Yashodhra Bajoria & Neha Goenka	CAXpert	Accounting Services: online platform to bridge the gap between CA and business owners

women entrepreneurs not only fill the economic gap of the nation but also inspire other women, solve the problems of several pertinent women entrepreneurs and generate job for other needy women. As per the Tenzin Norzom, June 15, 2020, women who are empowering SME in India through their services. (See Table 7.4 above.)

7.11 CONCLUSIONS OF THE STUDY

In the present world, women are recognized as invisible participants to strengthen the stable economy globally, and the same case in India women became the change symbol for society, economy, and importantly for themselves. Their participation in both formal and informal sectors has increased significantly in the past decades. Society and government also noticed the role of women in development of the nation and realized that there is need to take steps for the promotion of woman entrepreneurs. Women have shown their capabilities in the business world and government schemes, and other institutional supports have proven the biggest facilitator to promote entrepreneurship among women. Factors that hinder the development of women entrepreneurs are not in terms of finance and infrastructure but to make a platform for the entrance of woman entrepreneurs. Despite all the best efforts made by the government, the concept of women entrepreneur has not fully flourished, reason being is the acceptance and trust has not developed of society toward women. To overcome this, need to be accepted that women are not less efficient than men. We have to remove the mindset that women should be confined to the homes and for secondary roles and bring them out so that they can be independent, self-reliable, self-respecting, and enterprising too. Only women can break their images to come out and show their capabilities to do something exceptional out of there so-called image. Nowadays, many Indian women became a successful through fueling their entrepreneurial spirit and providing necessary support to women for their establishment.

7.12 RECOMMENDATIONS OF THE STUDY

This study has shed light on the facilitators and barriers of woman entrepreneurship in India. On the basis of review of literature, the study comes out with a few suggestions for the flourishment of woman entrepreneurship such as:

1. The government should design a specific development program for women by considering them a specific target group.
2. Sincere efforts need to be implemented in the direction of women education.
3. Training and development program for new women entrepreneurs.
4. Easy steps for buying and selling of raw material for women entrepreneurs.
5. Banks should also work on a special cell working on to providing finance for women entrepreneurs.
6. Need to create a favorable environment for women to support women to get into entrepreneurship.
7. Modernization of society is must so that they accept women also as an important participant in the economic development of nation.
8. Women education also plays an important role in the development of women entrepreneurship.
9. The role of NGOs and women entrepreneur institutional bodies also build the strong platform for women entrepreneurs.
10. Young women entrepreneurs need to get much more financial and moral support from the environment.
11. To make aware women about the digital business and world.
12. It is more psychological to flourish women entrepreneurs in India. For that, it is required to make discussion on present case studies of existing women entrepreneurs help to inspire women to choose business career.

7.13 IMPLICATION OF THE STUDY

Research Implication: This is study is helpful to further conduct study in this field with more specific variables, region, and case studies. Further studies can be conducted from the prospective of urban and rural entrepreneurship. More studies need to be done to know the behavior of women toward entrepreneurship, so this study can be the base to find out the variables for the same.

Practical Implication: It will be helpful in to giving awareness about women's entrepreneurship status in India. The study has shed light on the government schemes and policies, so concern bodies may get benefited from this. The researchers have discussed some successful entrepreneurs, which may motivate other women to start their own trade to generate their living and create for others. Suggestions provided by the study give an idea to the government, NGO, and institution to work on it for further development in women's entrepreneurship. This study can be the base to talk about women entrepreneurship.

7.14 LIMITATIONS OF THE STUDY

This study is a qualitative study based on the review of available literature. This study has covered only limited objectives. The researchers have generalized the study but not focused on a specific group, region, or attributes.

REFERENCES

Agarwal, S. (2020). Gender differences in quality of work life: an empirical study. *International Journal of Knowledge-Based Organizations (IJKBO)*, 10(4), 52–59 (ISSN: 2155-6393).

Anggahegari, P., Yudoko, G., Rudito, B., Mulyaningsih, H. D., & Ramadani, V. (2021). Female social entrepreneurship in Indonesia: a critical literature review. *International Journal of Entrepreneurship and Small Business*, 43(3), 315–331

Dana, L. P., & Dana, T. E. (2005). Expanding the scope of methodologies used in entrepreneurship research. *International Journal of Entrepreneurship and Small Business*, 2(1), 79–88.

Devi, P. (2013). Facilitating factors for women entrepreneurship in India. *International Journal of Science and Research (IJSR)*, 4(5), 153–156.

Garg, S., & Agarwal, P. (2017). Problems and prospects of woman entrepreneurship–a review of literature. *IOSR Journal of Business and Management*, 19(1), 55–60.

Goyal, M., & Prakash, J. (2011). Women entrepreneurship in India-problems and prospects. *International Journal of Multidisciplinary Research*, 1(5), 195–207.

Idris, N., & Tan, J. (2017). Review of literature on women entrepreneurs. *Journal of Global Business and Social Entrepreneurship (GBSE)*, 1(3), 1–11.

Jasuben ka pizza: Narendra Modi takes a jibe at Rahul Gandhi, media. Daily Bhaskar. 8 April 202021. Retrieved 8 April 2021.

Kaur, A. (2018). Women empowerment through five-year plans in India. *International Journal of Advanced Educational Research*, 3(1), 42–45.

Kaushik, S. (2013). Challenges faced by women entrepreneurs in India. *Education*, 35(53), 53.

Kaviarasu, S. J., Hendry Ruban A., & Francis, C. (2018). Women entrepreneurship in Indian context: A critical study of its challenges and solutions.

Khaire, R. J. (2011). Literature review of the women entrepreneurs and statutory policies. *ZENITH International Journal of Multidisciplinary Research*, 1(7).

Kumar, S., & Gupta, K. (2013). Social entrepreneurship: A conceptual framework. *International Journal of Management and Social Sciences Research*, 2(8), 2319.

Manolova, T. S., Brush, C. G., Edelman, L. F., & Elam, A. (2020). COVID 19 Pivoting to stay the course: How women entrepreneurs take advantage of opportunities created by the COVID-19 pandemic. *International Small Business Journal*, 38(6), 481–491.

Ojong, N. & Simba, A. & D, Leo-Paul, (2021). Female entrepreneurship in Africa: A review, trends, and future research directions. *Journal of Business Research*, 132(C), 233–248.

Ramesh, B. (2018). *Problems and Prospective of Women Entrepreneurship in India.*

Rana, G., & Sharma, R. (2017). Organizational culture as a moderator of the human capital creation-effectiveness. *Global HRM Review*, 7(5), 31–37.

Rana, G. & Sharma, R. (2019). Emerging human resource management practices in Industry 4.0. *Strategic HR Review*, 18(4):176–181.

Saraswat, R., & Lathabahvan, R. (2020). *A Study on Women Entrepreneurship in India.*

Sharma, R., & Rana, G. (2021a). An overview of entrepreneurship development programs in India. *International Journal of Environmental Policy and Decision Making*, 3(1), 16. https://doi.org/10.1504/ijepdm.2021.119511.

Shivani, A. (2021). Impact of demographic variables on subjective wellbeing of students. *Journal of Public Affairs, Web of Science (Clarivate Analytics) and SCOPUS (Elsevier), Emerging Sources Citation Index (Clarivate Analytics) Vol. X, No. Y, xxxx (Scopus, ABDC Listed) (1479–1854) (Citescore 1.1).*

Singh, A. (2014). Role of women entrepreneurs in India: A SWOT analysis. *International Journal of Management and International Business Studies* 4(2), 231–238.

Sivanesan, R. (2014). A comparative study on rural and urban women entrepreneurs– Prospects and challenges. *International Journal of Research in Management & Business Studies*, 1(3), 28–34.

Yadav, V., & Unni, J. (2016). Women entrepreneurship: research review and future directions. *Journal of Global Entrepreneurship Research*, 6(1), 12.

8 Role of Business Incubator and Innovative Practices in Entrepreneurial Development

Pushpendu Rakshit
Amity University

Ashwini Kumar Saini
G. B. Pant Institute of Engineering and Technology

CONTENTS

8.1 INTRODUCTION

As homegrown Indian organizations and western companies working across India take a stab proceeding with development, they have a lot to acquire from taking on and rehearsing these six administrators such as looking for a promising circumstance in difficulty, accomplishing more with less, thinking & acting adaptability, keeping it basic, including the edge, and following the heart. Organizations working across India should perceive that the organized model of development which ruled the 20th century is similar to a symphony – top-down, inflexible, and driven by upper-level representatives. These models functioned admirably in the stable universe of

DOI: 10.1201/9781003217084-8

abundant assets. However, at times given the complicated and unusual business conditions that the organizations are going to look in forthcoming years to boost their business immunity. They also require an elective methodology, which is similar to that of a jazz band-based improvisational, liquid, community, within a casing work of profound information. Inventive practices in business addresses the other option. But keeping in mind that innovative activities may entail a variety of challenges over time. Impromptu game designs can work on occasion, but not all of the time.

Western pioneers acclimated to a universe of relative plenitude and were used to working for such a long time in a somewhat unsurprising climate. Might be the experiences assist betterment in practices and embrace from developments of past. Trend-setters find open doors in misfortunes in three ways: rethinking difficulties as happenstance as development, making limitations work for themselves, and continually embracing a changing climate by ad-libbing answers for difficulties they face in route. The first, and maybe the main procedure of pioneers, is rethinking or changing the focal point through which they see the circumstance they face. Trend-setters see and decipher the world uniquely in contrast to most of us. Their capacity to re-examine implies they are probably going to see the glass half full in any event, when each one considers it to be half vacant. One might consider jugaad pioneers as current chemists who intellectually change misfortune into a promising circumstance. However, looking for a promising circumstance in misfortune is just the primary the numerous development administrators that the corporate pioneer and their associations can apply to prevail in an eccentric and quick worldwide economy. None of the numerous types of difficulty that emerging economies and firms face is as severe or as detrimental to endurance and development as scarcity.

Pioneers can get more from less by applying thriftiness to each action they perform at each progression along the worth chain. They are economical by the way they plan items, how they assemble them, how they convey them, and how they perform after deal support. Their moderation is evident not just in their stingy utilization of capital and regular assets, but additionally by the way they augment their restricted time and energy: rather that doing everything without help from anyone else, they broadly depend on accomplices to perform different tasks, hence saving time and energy. Coming up next are some of economical methodologies utilized by pioneers to acquire more with less. Enormous organizations should perceive that the developments are not just with regard to planning reasonable items. It is likewise about effectively selling these items in commercial centers. However, fruitful selling will not occur if selling individuals are motivated to sell only first-class things. All things being equal, organizations should adjust their business power impetus framework to the corporate methodologies of accomplishing more with less. Organizations can resolve this issue by perceiving their sales force along band lines, with various sales reps liable for low-end and very good quality portions. This will likewise assist with lessening any inside obstruction dependent on the dread of cannibalization. Far superior, sound inward rivalry between divisions could drive deals and showcase work force liable for contrast brands being more imaginative by the way they reach and keep their particular clients.

There are a lot of benefits received from incubators for business start-ups. Also, lot of infrastructural support along with staffing resources are made available at very reasonable rates as compared to the market. From funding to planning, there is a similar level of support provided by the incubators. There exists a need of strategic support for ideas to nurture further. For any new business, they almost spend around 2 years in an incubation time frame where they are provided with infrastructural and technological assistance along with facilities such as telephonic, equipment expenses, internet access with other start-up companies to control operational costs (Sharma and Rana, 2021). Thus, it's understood that business incubation is a place provided to the naïve businesses, facilitating them with all the resources needed at initial stage of business all under one roof. This would also assist the entrepreneur to utilize maximum time of development process rather than spending time on locating resources. In addition to this, other than making provision for desk or office incubators, they often provide resident assistance along with expert guidance, legal advisors, mentors, office equipment, gadgets, technological support, training, and opportunity to meet with investors and other stake holders (Ravindra et al., 2021). Thus, incubators are considered as launching pads for new businesses who look forward to having a high fly in the world of business. This appears to be an institution that supports them to nurture and eventually graduate and become successful business czars. However, due to certain unavoidable uncertainties, some are unable to be successful participants and some fail to shape their dreams/ideas which lead to shutting down instead. There are many hurdles one must be ready to cross when it comes to the success of a business. It might be that just one is very close to their mission success when they thought to quit.

Numerous written works contribute to creative practices (Tushman, furthermore Moore, 2005), keeping normal linkages noticeable. Drucker (1985), in his view, expresses advancement as training that exhibits acknowledgment of corrections in business situations on numerous occasions. Myers (1996) divides curiosity into two phases: stage one, where people carry out a "gathering of exercises completed at miniature level" in logical inconsistency of "a bunch of large-scale infrastructural situations" that immaculately "energize, make loose, and further develop oddity" at the miniature stage, on the other hand, outside natural highlights additionally endorse a first piece into play, and regardless of its latent capacity, oddity doesn't, in general, append with business achievement (Jorna, 2006). Kandampully (2002) communicates that advancement is must for center turns of events and supportability in any business space. It's the capacity and innovativeness of an individual to improve and start guileless strategic approaches as a pioneer, money manager or trough they should hold (Sharma et al., 2021a,b). New organizations carry benefits and opportunities for others to get utilized, subsequently contributing in friendly government assistance (Rana and Sharma, 2021). It's additionally perceived that this would doubtlessly be a difficult errand and that one should keep while wanting to begin a new pursuit (Shlomomaital and Seshadri, 2007). Curiosity abilities are only innovativeness. Cultural therapist Mihaly Csikszentmihalyi (1996) talks about the presence of remarkable characters in the fields of workmanship, writing, and science. He brings in his theory of imaginativeness as "stream" creativity. It doesn't appear inside individuals' brain; however they see as a result of cooperation between an individual's perspectives and social-social structure. Creative practices are the greater part of the most ideal ways to exhibit thoughts and manners of thinking according to the needs of time and mankind. Such advancements are, for the most part, found in assembling and medical care organizations, Lyons et al. (2007), redirecting consideration toward the need for improvements according to social and social requirements. Organizations must occasionally present advancement, either supported by hatcheries for new participants or by themselves without the assistance of anyone else (Rana et al., 2019). Hendry et al. (2008), say that modern foundations involve more stately techniques for development for victories when compared to service foundations.

8.2 PROGRESSION IS MORE THAN CREATING: IT'S ALSO A CRITICAL PROCESS

Schumpeter (1934) welcomed a few debates on how affiliations look for new freedoms and strategic positions. One of his most renowned debates has to do with the chance of "inventive pulverization". In Schumpeter's vision, a private undertaking is a dynamic and innovative market structure that is reliably progressing. These changes, he says, happen considering essential examinations and cycles that acknowledge control over the current point of view (the norm) and further develop approaches for passing on things, associations, or new pursuits. The cases of annihilation and irregularity are what make private undertaking make and make, and not the generally perceived thought of monetary concordance. Thusly, the Austrian financial analyst alludes to this awkwardness as "imaginative decimation", since progress as regularly as potential prompts market issues: for example, things go downhill, affiliations close, and individuals get jobless. Considering everything, he recognizes these are essential ramifications for new strategies that advancement life quality to become evident.

We've been watching these instances of imagination crushing across mankind's plan of experience. We've gone over the central current change to the effect of utilizing oil, and today we are an especially imaginative society of data. These periods of progress make the old models out of date and make the essential for everybody and everything to make – and the change is phenomenal. A business point of view, only relationships with a phenomenal urgent vision continue on through a scene of innovative annihilation. Companies such as Nokia and Kodak, which were once market leaders in their respective fields, failed to anticipate how the cellphone and photography experiences would change. Simultaneously, from an agreeableness system, as we've investigated in this article, affiliations making moderate improvements and growing genuine sensibility strategies (from managing waste in a round method for utilizing attainable power) will likely be the ones flourishing later on.

8.2.1 Development and the System's Perspective

As demonstrated by the Oslo Manual, the structural perspective of improvement is connected to checking out the interdependencies among performers. As the progression method has problematic outcomes and follows an awesome and non-straight technique, the structure perspective is a multidisciplinary and interdisciplinary one. This suggests advancement systems can be related to industry, geography or development and are regularly interrelated.

According to the OECD, structural perspectives are used to cause advancement game plans to coordinate system changes in the public eye. From the decarbonization of transportation structures to the improvement of the guidance system in showing coding, these progressions require different performers, from producers to purchasers, to participate and ensure that all of the comparing portions of a baffling association are set up. For instance, following Oslo Report's model, for creators to encourage their extent of electric vehicles, more people need to buy from this section, and an association of charging stations ought to be made as well.

8.2.2 Different Curiosities about the Process of Innovation

Advancement is a persistent learning process with a few sources of info. It's anything but a direct or successive cycle, as it includes numerous connections and criticism. A model or thought of an item that doesn't yet exist is overall not an item advancement as it doesn't meet the execution necessity. This state of placing into utilization or making accessible for the utilization of people, firms, or different associations is the thing that separates development from different ideas like "creation". The monetary and social effects of thoughts and inventive changes rely upon the dissemination and adoption of related advancements.

8.2.3 Business Incubators – A Handholding to Start-Ups

A business hatchery is a work environment formed to offer new companies and new undertakings admittance to the assets they need, all under one rooftop. This brings us to one point, representing the growing business visionaries, where they can enjoy the execution of arrangements that are meant to rejuvenate the business. The term "hatchery" fundamentally demonstrates that there should be support at the starting stage for any business to develop and sustain, as a large number of the point financial backers even anticipate contributing to and contributing to the betterment of business. Numerous hatcheries are from explicit businesses, and many acknowledge that they are even from different enterprises. There have been numerous models, for example, Ola, Uber, Zomato, Yulu, and Swiggy, which have been doing great in business as an adult business. Additionally, as of late, NYKA has demonstrated the same by getting recorded in the market with its IPO. There have been numerous chances for some organizations as of late where they have developed well, and surprisingly, many have contracted with time. Some specific hatcheries take into account education, FinTech, or monetary innovation, ceramics, green innovation, homeland security, fashion, and food. A large number of the hatcheries and point financial backers anticipate support either by all means or just monetarily to help the resolve of business. For example, Anand Mahindra recently declared that he would provide financial assistance to one of the young entrepreneurs who wished to have sports shoes made out of plastic waste. Consequently, there are numerous such outlines that show that organizations have searched for help for their endurance and manageability.

Throughout the long term, specialists have characterized business incubation in their specific manner. The fundamental idea, notwithstanding, continues as before. As per Sherman and Chappell (1998), a business hatchery is a "monetary improvement device fundamentally intended to help make and new organizations locally." Further, Sherman and Chappell (1998) note that the business hatchery upholds arising organizations with a few administrations, for example, help with building supervisory crews, creating business and advertising plans, reserves, proficient administrations, shared gear and that's just the beginning. There exists a shortage of assets for the business visionaries yet they have thoughts that are rich naturally. More often than not it's been seen that there are ways by which one needs to utilize assets. Hatcheries are lifesaving emotionally supportive network for youthful growing organizations. The majority of the organizations even depend on point financial backers for ventures. Accordingly, it's perceived those main thoughts are not adequate but rather additionally one necessity to find assets to rejuvenate it. The greater part of the business in the contemporary situation is seen with key moves, for example, Oyo rooms, where the author doesn't own any room. Indeed, even if there should be an occurrence of Ola and Uber as vehicle rental specialist co-op, yet don't own any vehicle. There are a lot of new companies with creative thoughts and in this way, there exists a need for even imaginative key methodology for the business. In this way, they require business hatcheries to perform critical jobs or fill the hole.

The following are the services offered by the incubators:

- Help a business monetarily and functionally.
- Help identified with marking and advancement is accommodated new businesses.
- Business hatcheries have a solid organization and network supporting the organizations to develop and associate across topographical limits.
- They likewise give help to showcasing research and special exercises to build perceivability.
- Help is accommodated overseeing bookkeeping books and accounting.
- Hatcheries carry dependability to the business. This guides the organization to get credits and monetary conveniences from monetarist foundations.
- Frequently the new companies don't have a clue how to make a compelling show to intrigue private supporters, funding, and different financial backers. Business hatcheries, with a lot of involvement behind them, assist these organizations with the introductions also.
- Hatcheries additionally goes about as a tutor, guide, and teacher to prepare and foster authority fitness among new pursuits.
- They additionally give availability offices and direction in regards to advancing, dispatching an item alongside finding enhanced business opportunities.
- They are additionally furnished with special and marking support.

8.3 INNOVATIVE ENTREPRENEURIAL PRACTICES

A business must adopt innovative practices to sustain business in a contemporary scenario. Innovation is the key to success and must be an integral part of any business owner. There must be a business process reengineering in every business. Innovation is an integral part of business cosmos and thus must be implemented properly. There are also instances recorded where the improper application of strategies might prove harmful for businesses. Thus, one needs to be very careful while adopting innovative business practices.

Looking for development procedures will regularly uncover tremendous amounts of something essentially the equivalent, or comparative, imagination practices as we've examined. To go past ideation works out, we will wrap up with an establishment of reasoning that can help when you are dealing with a wide extent of progressive issues. Essentially, open headway joins looking for and tracking down plans outside of the different evened-out arrangement. Open improvement is really hard to make sure about. The instructor and producer Henry Chesbrough was one of the first to depict it: "Open movement is 'the use of purposive inflows and outpourings of information to speed up inside progress, and extend the business regions for outside utilization of headway, respectively.'" Six with everything considered, affiliations subject to an arrangement of open improvement look past their creative work abilities to manage issues. This viewpoint can facilitate a wide extent of thing and association movement processes. Open improvement models also award movements to be shared all things considered so they can seed different headways outside the principal firm or foundation.

Open progression takes a certain perspective on sharing data and thoughts across a general populace related by quick correspondence affiliations. It is in like the way a shift from the superb creative work model. It could be said, you permit others to manage issues in your business, fire up, or social business project. In this relating world, you are free to the reality that data are hard to leave stowed away. You might look for licenses for your got improvement, especially in fixed thing or association practice structure, however, you ought to expect, or even invigorate, the broad dissipating of key pieces of your reactions. This looks great: If, as a monetary-trained professional or a creative endeavor, you will look past your ideation, appraisal, and movement limits regarding plans, you should expect that others will search for your reactions for contemplations to get.

The open improvement model is certainly clearer to depict in sure terms than it is to endeavor without moral outcomes. Incredibly, current and corporate perception, thievery of guaranteed improvement, and cases are customary. Considering everything, motivation in movement can emerge out of heap sources when predictable floods of data are free to anybody with a fast information association. Open improvement is an unmistakable yet critical development for future movement and for overseeing, even perhaps planning, irritation in an industry as examined early (i.e., perilous progress).

Another part of the open progression model is the relationship between academic investigation and practical game plans. The reciprocal effect between the insightful local area, which as often as possible moves step by step, and driving corporate and ambitious powers, which consistently revolve too scarcely around transitory increments, could offer the balance this rapidly changing world necessities. In case you can sort out some way to plug into the exchanging of musings between longstanding establishments and dangerous mechanical pioneers, you may be arranged to affect positive change in society and to encourage things that are gotten as supportive and dazzling, amazingly new and inventive, and vital for the human association with a comparable time.

Table 8.1 shows the list of few companies who contributed effectively in the adoption of disruptive technology.

TABLE 8.1
Illustrations of Disruptive Technology

Company	Disruptive Technology
Amazon	Speed based delivery Multiple delivery processes from drones to strategically located fulfillment centers Disruptive technology includes processing the customer order before the customer has even finished the purchase so that the product is already moving toward delivery
Uber and Lyft	Ridesharing versus taxi driving Apps and Beacon and Amp-color-coded alert communication system disrupted the taxi system
Bitcoin	Digital currency is not connected to a specific country or monetary standard Value is based on market forces
Toyota E-Palette	Remote-controlled driverless electric shuttle that brings the service to the customer rather than the customer going to the service

8.4 STEPS FOR OSMOSIS OF IMAGINATIVE STRATEGIC APPROACHES

- Use Technology to Collaborate and Share Knowledge. Joint effort drives ingenuity and improvement, and online media and conferencing degrees of progress can assist with joining individuals (or in every practical sense, together) more a large part of the ideal opportunity for that cooperation.
- Advance Innovation as an Organizational Value. The most imaginative affiliations didn't only karma into enrolling creative individuals; they set inventive and amazingly commonplace individuals into innovative social orders.
- Join Innovation as a Leadership Development Competency. Part of building an innovative culture is having pioneers who respect creative minds, and are imaginative themselves.
- Append Compensation to Innovation. The jury is now thinking about the impact of impetuses on inventiveness, however, their use in affiliations passes on a message that movement is respected. That sign is a gigantic piece of culture building.
- Make a "Thought finding" Program. As we've examined somewhere else, it's insufficient to have exceptional insights. Imaginative affiliations foster a framework that takes advantage of the absolute information on everybody and licenses everybody to impel sagacious insights.
- Store Outside Projects. It may sound unusual to permit financing to empower tries that are believe it or not outer your connection, however as market limits proceed to dim, key progression affiliations become impressively more immense.

- Train for Creativity. The imaginative psyche isn't normal. Imaginative reasoning limits can be made and the most innovative affiliations store preparing exercises to empower them.
- Make a Review Process for Innovative Ideas. Even the best thoughts don't come from totally experienced. There is a cycle to refining, making, and seeing the insights with the most market potential. Making an audit cycle permits this to occur and hails those imaginative contemplations are respected.
- Enroll for Creative Talent. Particularly at the undergrad and graduate levels. The battle for limits is moving back moving its obsession from quantitative characteristics to innovative ones.
- Reward Innovation with Engaging Work. Examination shows that affiliations that can perceive their most innovative agents can additionally foster their inventive cutoff by giving them the independence to oversee projects that are regularly enthralling to them.

8.5 DEVELOPMENT TYPES

Thing improvement suggests new or by and large further evolved things. The improvement could join making the very thing with better parts or another clearer thing.

Cycle headway suggests new ways a business can pass on its thing or organization. This could recall changes in techniques, equipment, or programming.

Promoting headway has to do with the way wherein the thing or organization appears to customers. It fuses new packaging or advancement campaigns, yet it moreover fuses any worth changes.

Legitimate headway implies higher viewpoint changes that a business can endeavor to additionally foster execution. This could be an internal change to key arrangements or an external opportunity to business relations with various substances.

The movement respect chain offers a reattempted and worked with strategy for overseeing evaluating association's improvement execution and figuring out which of the many recognized frameworks out there would be ideal to take on. The chain-based view can assist supervisors with conveying a surge of new things and associations. Considerably more all around, it can assist them with at long last understanding the potential from their headway theories.

The improvement respect chain can in addition assist bosses with the understanding that a clear progress strength may genuinely end up being a lack when chiefs target gives over the most grounded partakes in the movement respect chain seeing remarkable bearing for supporting a center breaking point in, say, thought age or dissipating they regularly further debilitated individual the feeblest pieces of the chain, undermining their headway limits when in doubt. To uncover new measures and occupation in the progress association, the supervisors tailor their answers for the right issues, over the long run, a failure point in the improvement respect chain will change into a solid one and one more piece of the chain will require thought considering everything. Directing body ought to be depended upon to isolate each affiliation the chain reliably to on and on work in the ordinary course of chain. Mark of truth, for firm bosses who embrace the progress respect chain viewpoint, it's not only the same old thing.

8.6 POINTS TO AVOID WHILE COMMENCING WITH START-UP

1. **Unstable Cash Flow**

   The endurance of any business is subject to predictable income. While slow income is normal in the first place, if it keeps on being an issue past the underlying dispatch time frame, the business will battle until the circumstance is amended—or the endeavor falls flat.

   There's a popular line from the 1989 film Field of Dreams: "In the event that you construct it, they will come." The line has acquired a foothold in the years since however, this reasoning can't be applied to a business start-up. The following several motivations behind why:

   All together for a business to be fruitful, there must be an exhibited market for the item or administration. If you don't test-market your item or administration before officially carrying out your business, you'll have no clue if it will even work ahead of time.

   If market-testing demonstrates the presence of a client base that will pay for your item or administration, then, at that point, the probability of achievement will be a lot higher. It'd then, at that point, involve increasing your activities when the business is officially dispatched. The option is dispatching your business at max speed while at the same time attempting to distinguish and offer to a specific market portion (not great). On different occasions, income issues are the aftereffect of a powerless advertising program. In a perfect world, you're promoting and your items/administrations will be test-advertised simultaneously. In any case, one of the issues with advertising is that it very well may be a significant expense and unsupported by Return on Investment (ROI) first and foremost – and the best way to manage that is to have sufficient accessible cash flow to finance your showcasing tasks.

2. **Excessive Overhead**

   With regards to working costs, utilizing a sound measure of moderation is fundamental. In any case, it's surely not true for numerous new businesses. You've heard the renowned saying: "Assuming you need to bring in cash, you need to go through cash."

   While there's a reality to that expression, numerous new organizations get excessively out of hand. Keeping your costs at a flat-out least is a significant procedure for guaranteeing the endurance of your business. Ideally, your start-up costs might be the no-frills least expected to get the entryways open.

   The thought isn't such a great amount to waste no time directed toward the beginning, but instead to develop into the business as income extends and legitimizes a higher spending plan. The accentuation ought to be on coordinating capital into showcasing and protecting money for future development. In case you're paying a lot for lease, pay rates, obligation administration, and hardware leases, you might have made a primary overhead your start-up can't uphold.

   All things considered, dispatch your business from home or investigate office space sharing. Recruit outside project workers dependent upon the situation rather than employing representatives (Sharma et al., 2021a,b).

Buy used furniture and hardware. Fundamentally, keep away from obligation no matter what!

3. **You've Got a "Me Too" Product/Service**

   One more explanation numerous new companies keep on battling is on the grounds that the item or administration does not have a remarkable selling point (USP). What's more, shockingly, this issue is regularly outlandish.

   To fill a beneficial specialty, your item should either be superior to the opposition's or similarly as great yet more affordable. In the event that you can't influence either front, no doubt about it "me as well" item/administration. Except if you're ready to work on quality or lower your costs, future possibilities are faint – you might need to return to the planning phase and make another item/administration dependent on a convincing USP.

4. **No Viable Processes or Systems in Place**

   Since your business is new, you probably won't realize what cycles or frameworks you really want. Be that as it may, their nonappearance can leave expanding openings in your strategy, and will not go unrecognized by your demographic. For instance, possibly you can come up short on intelligent records receivable capacity which brings about a huge slack between installment assortment and item conveyance.

   As a start-up, you can concentrate on related organizations and see what cycles and frameworks they have set up. You can likewise consider going straightforwardly to your customers and asking them what upgrades could be made!

   You'll have to address these right off the bat in the start-up interaction as could really be expected. The littlest of issues have a method of becoming ongoing as the business develops – so, all in all, executing new frameworks can turn out to be restrictively costly, totally troublesome, or both.

5. **Assembling the Wrong Team**

   You without a doubt carry specific abilities to your business, however, it's improbable that you have each expertise expected to carry your start-up to progress. For instance, perhaps you're the inventive advertising virtuoso behind the brand, however, you need managerial, client support, or direct deals abilities. You'll need to recruit individuals with these abilities together for your business to turn into a full-specialist organization.

   We understand this is a difficult task since income is a regular issue among new businesses. You most likely won't have the assets to enlist a full staff that can give every one of the administrations expected to make your organization work, especially assuming your representatives need regular work and the manager-given benefits.

   Another option (basically during the start-up stage) is to either subcontract portions of your business to outside suppliers or to recruit laborers on an agreement premise. This will assist you with keeping your finance at a flat-out least since you'll just be paying for administrations dependent upon the situation.

   Clearly, neither one of the choices is great. Reappropriating may bring about your business coming up short on that consistent quality, while contract representatives might be reluctant to work on low maintenance or with a variation pay. All things considered, you'll need to sort out a mix of the two that best suits your requirements until you're in a situation to enlist ordinary representatives.

6. **Partnering with the Wrong Investor**

   At the point when you're dispatching a start-up, it's enticing to take cash from any source ready to give. In any case, thusly, you might wind up in the appalling situation of managing a financial backer who doesn't share your vision, or who attempts to apply impact in pointless ways. This is particularly evident if the venture contract provides the financial backer with any degree of command over everyday activities.

7. **Lack of Flexibility**

   Energy and fervor about the assistance you bring to the table for the world is fundamental – however, it can bring about limited focus. Numerous new companies have bitten the dust on the grounds that their authors were so enamored with their own thoughts that they wouldn't see the requirement for changes. Keep in mind that individuals purchase the things they think they need, not the things you think they need.

8. **Failing to Identify Customer Persona**

   Making a viable promoting plan a fundamental stage for your start-up. In case it isn't done well (or done by any means), it'll be difficult to acquire an underlying client base. The best-promoting endeavors are those designated at the gathering that (1) has the greatest requirement for your item or administration and (2) is in the best situation to buy it. If you neglect to distinguish an optimal client persona, your advertising will be excessively nonexclusive, and in this way inadequate.

9. **The Wrong Location**

   For most new businesses, cash is tight, and it can appear to be savvy to dispatch your store in an area where lease and utility expenses are really low. Sadly, there's frequently a justification for why rates are low in a space – helpless foundation, absence of traffic, absence of motivating forces from neighborhood legislatures, and so forth ensure you research your designated topographical area to foresee whether or not there's a nearby market for your item!

10. **Completely Refusing Debt**

    Obligation can be unnerving in any event, when you're procuring a benefit, so collecting obligation when you're not bringing in cash can be out and out alarm initiating. Lamentably, except if you're as of now loaded from different endeavors, straying into the red is essentially important for dispatching your start-up.

    Getting a decent credit extension and a business advance will permit you to buy stock, cover month-to-month costs, pay opening expenses, and keep you above water!

11. **Giving away too Much Equity**

    While there's nothing innately amiss with giving financial backers a value stake as a trade-off for their subsidizing, parting with a lot of can create issues. Parting with value is parting with control, and regardless of whether you're giving it out in little augmentations, it collects across various financial backers. Bedlam can follow if your financial backers disagree regarding what should be done to deliver development, and more terrible – you may wind up unexpectedly having little impact on your own start-up.

12. **Building a Team Based on Friendship**

    Collaborating with loved ones to dispatch a business is a great dream. Who wouldn't have any desire to go through their days working with individuals they like/love the most? Tragically, the outcomes are frequently not generally so blushing as you'd suspect. It very well may be hard to explore debates and characterize individuals' jobs and obligations when managing loved ones!

13. **Not Enough Investment Funding**

    Numerous new companies battle or come up short since they just score sufficient subsidizing to begin – they don't take development, market movements, or expected calamities into thought. Then, at that point, when something sudden occurs, they're too caught up with scrambling to get the assets they need to zero in on what's more significant: showcasing, recruiting, and item improvement.

8.7 CONCLUSION

The headway in the business communication makes the things isolated that can bring liberal additional value, relative to what customers can buy at this point can succeed. To achieve this, affiliations/firms should have better capacities, abilities, and significance being developed communication and its organization. In the present universe of overall business areas, the associations which can progress and manage advancement will be the survivors who will be through and through better contrasted with their adversaries. Further, in the value chain, pattern of innovation change depends upon various grouping of cutoff points of the associations that are capable for advance toward overseeing steady-state and interminable turn of events. Consequently, the players of the huge worth chain have various choices between these two farthest places, including setting up remarkable units inside setting up business affiliations, or to encourage administering more "open turn of events" assignments which will help the ambitious characteristics of little players in regard chain. Thusly, the significance of advancement in overall worth chains is basic for shrewd how firms in different bits of the world can develop permission to general business areas, to assess the possible benefits and the perils inferred meanwhile, and how to grow the net increments from responsibility in overall worth chains can be improved and strengthen. The methods of viable improvement in the overall economy are an ordinarily intriguing and insignificant objective, our goal is, in general, to help by having a reasonable sensation of the advancement cycle in various ways of critical worth chains and the key determinants that graph these effects.

REFERENCES

Mark H.Moore (2005). "Break Through Innovations and Continuos Improvement : Two Different Models of Innovative Processes in the Public Sector", Public Money & Management, Taylor and Francis Journal, Vol 25(1), pages 43–50.

Hendry, L., Huang, Y. and Stevenson, M. (2008). Workload control: Successful implementation taking a contingency-based view of production planning and control. *International Journal of Operations & Production Management*, 33(1): 69–103. https://doi.org/10.1108/01443571311288057.

Influence on Entrepreneurship (and Management) Research, Schumpeter's (1934), *Revista de Empreendedorismo e Gestão de Pequenas Empresas*, 6(1): 04–39. Jan/Apr. 2017, doi:10.14211/regepe.v6i1.483.

Jorna, R. (Ed.). (2006). Sustainable Innovation: The Organisational, Human and Knowledge Dimension (1st ed.). Routledge. https://doi.org/10.4324/9781351280365. "https://www.emerald.com/insight/search?q=Jay%20"

Kandampully, J. (2002), "Innovation as the core competency of a service organisation: the role of technology, knowledge and networks", "https://www.emerald.com/insight/publication/issn/1460-1060" European Journal of Innovation Management, Vol. 5 No. 1, pp. 18–26. https://doi.org/10.1108/14601060210415144.

Lyons, B. D., Velada, R., Caetano, A., Michel, J. W., and Kavanagh, M. J. (2007), The effects of training design, individual characteristics and work environment on transfer of training. International Journal of Training and Development, 11: 282–294. https://doi.org/10.1111/j.1468-2419.2007.00286.x.

Moneta, G. B. and Csikszentmihalyi, M. (1996), The Effect of Perceived Challenges and Skills on the Quality of Subjective Experience. Journal of Personality, 64: 275–310. https://doi.org/10.1111/j.1467-6494.1996.tb00512.x.

Rana, G., & Sharma, R. (2021). Employer branding: attracting and retaining employees for sustainable development in disruptive economy. *World Review of Science, Technology and Sustainable Development*, 17(4), 319. https://doi.org/10.1504/wrstsd.2021.117839.

Rana, G., Sharma, R., Goel, A. K. (2019). Unraveling the power of talent analytics: implications for enhancing business performance. In Rajagopal, B. R. (ed.) *Business Governance and Society*. Palgrave Macmillan, Cham. https://doi.org/10.1007/978-3-319-94613-9_3

Ravindra. K & Suman Mor., (2021). E-waste generation and management practices in Chandigarh, India and economic evaluation for sustainable recycling, *Journal of Cleaner Production*, 221, 1 June 2019, 286–294.

Sharma, R., & Rana, G. (2021). An overview of entrepreneurship development programs in India. *International Journal of Environmental Policy and Decision Making*, 3(1): 16. https://doi.org/10.1504/ijepdm.2021.119511.

Sharma, R., Rana, G. and Agarwal, S., (2021a). Techno innovative tools for employer branding in industry 4.0. In G. Rana, S. Agarwal and R. Sharma, (eds.), *Employer Branding for Competitive Advantage*, 1st ed. Boca Raton, FL: CRC Press. https://doi.org/10.4324/9781003127826-11.

Sharma, R., Saini, R., Prakash, C. and Prasad, V. (2021b). *Internet of Things and Businesses in a Disruptive Economy*, 1st edn., New York: Nova Science Publishers.

Sherman, H., & Chappell, D. S. (1998). Methodological challenges in evaluating business incubator outcomes. *Economic Development Quarterly*, 12(4): 313–321. https://doi.org/10.1177/089124249801200403.

Shlomo maital and D. V. R. Seshadri, (2007), "Innovation Management': Strategies, concepts and tools for growth and profit, 2nd Edition, SAGE Publications Ltd., New Delhi.

Tushman ML, Rosenkopf L (1992). On the organizational determinants of technological change: Towards a sociology of technological evolution. Res Organ Behav 14: 311–347.

9 Technological Drivers and Entrepreneurship Challenges of the Manufacturing Sector in Industry 4.0

Identification of Critical Factors through Total Interpretive Structural Modeling (TISM)

Suha Bilquis, Afaf Akhter, and Ayesha Farooq
Aligarh Muslim University, India

CONTENTS

DOI: 10.1201/9781003217084-9

9.1 INTRODUCTION

The concept of I4.0 has brought a revolution in the business world due to its widespread applicability and strong potential of bringing innovation. Industry 4.0 is the result of continuous innovation through digitalization and the effective deployment of artificial intelligence (AI). I4.0 comprises properties of cyber physical system (CPS) and internet or intranet of things (IoT) which again assimilate the heterogeneous data and a knowledge base of different fields significant to the business environment (Sony and Naik, 2020). The continuous effort of humans and technology has led the business world from the First Industrial Revolution to the Fourth Industrial Revolution. The increased use of hydropower, steam power, development of machine tools, electricity, mass production (assembly lines), accelerated automation, and information technology has eventually caused businesses to follow CPS technology, integrating the real world with the information age for future industrial development (Zhou et al., 2016). Industry 4.0 originated through a German initiative, which provides a cross-section of different technologies rooted in mechanical engineering. There are no standardized rules or designs that restrict the evolution of Industry 4.0 (Bonaccorsi et al., 2020). Integration of digitalized technologies such as internet of things and services for delivering industrial value creation gave rise to this evolutionary paradigm referred to as "Industry 4.0" or "Industrial Internet of Things" (IoT) (Müller et al., 2018). Industry 4.0 came into existence because of continuous efficient efforts for eradicating problems of growing customization, reduced technology & innovation cycles, and better demand volatility (Vinodh et al., 2021).

Industry 4.0 Resolves Manufacturing Challenges: Features of Industry 4.0 resolve the challenges of 'e-operations', which refers to the collective integration and management of all the business operations, which can be done now with the help of enterprise resource planning (ERP) and IoT. Also, assessing data and making meaning out of it for smooth business operation (commonly known as big data analytics) poses a significant challenge. Implementing the right software for analyzing massive quantifiable data and, out of that, what to measure and how to measure & control resource usage is again a big challenge for manufacturing companies (Chiarini et al., 2020b). Other technologies of Industry 4.0, such as IoT and CPSs, involve synchronization within a business system as a whole and among different business plants to make the entire manufacturing process efficient with the least maintenance and breakdowns (Varghese and Tandur, 2014).

Barriers to Industry 4.0 in the Developing Countries: The lack of awareness among businesses related to different technologies associated with Industry 4.0, inadequate investment & infrastructure facilities such as wireless requirements, Wi-Fi chipsets, and lack of skilled competence are some of the major challenges restricting the scope of Industry 4.0 in a developing country like India (Stentoft et al., 2020; Varghese and Tandur, 2014).

9.1.1 Entrepreneurship in the Industry 4.0 Era

Practicing entrepreneurship in the Industry 4.0 era is a difficult task, especially in the developing countries as most of the entrepreneurs are nascent in these emerging economies, and there happens to be a misalignment between technology utilization and technology demands in terms of skills and capital (Zhou et al., 2016). Also, the deployment of technology and digital transformation within the business is a hurdle as sustainable issues come into the picture. New technologies and capital-intensive techniques mean higher energy consumption, which ultimately increases carbon footprints (Vuong, 2020).

However, entrepreneurial capacity is by and large central to all technological innovation, as only the right minds support investments in R&D, experimentation, creative thinking, and technological leadership. Specifically, evidence suggests that entrepreneurship is an antecedent to innovation in terms of product, technology, and market. Economies that promote entrepreneurial practices and invest in developing a proinnovation cultural background are more likely to become the leaders of industrial revolutions (Ferreira and Lisboa, 2019).

The following research questions are answered in this paper:

What are the Technological drivers of Industry 4.0 in the manufacturing sector?
What are the challenges of entrepreneurship in the manufacturing industries in the context of Industry 4.0?
What is the relationship between the identified drivers and challenges?

This paper is divided into two sections: the first section answers the first two research questions and identifies drivers/challenges of entrepreneurship in Industry 4.0 through a systematic literature review (SLR) using Preferred Reporting Items for Systematic Review and Meta-Analysis (PRISMA); The second section answers the third question by finding the hierarchy and relationship between variables by Total interpretive structural modeling (TISM).

9.2 LITERATURE REVIEW

To achieve the first two objectives proposed in this research paper, a SLR was carried out using PRISMA (Preferred Reporting Items for Systematic Review and Meta-Analysis) 2020 flow diagram. The SLR methodology was adopted because it summarizes the available literature in an orderly manner in response to specific research questions. The literature review began by defining the research questions, and related keywords, namely, Industry 4.0, entrepreneurship, technological drivers, and entrepreneurship challenges. Then, to search the articles related to these topics in the database, keyword combinations were performed.

After defining the keywords and their combinations, they were put in selected databases, namely Scopus, Web of Science, Science Direct, Springer, and Emerald. These databases were selected on the basis of the availability of access to the author on the Open Athens portal; other sources included search engines such as Google Scholar, and Research Gate, to further explore the literature (Figures 9.1 and 9.2).

The review of existing literature on the topics related to Industry 4.0 challenges and technological drivers after searching on several databases and search engines, with a timeline spanning from 2015 to 2021, resulted in 245 articles. Various keywords were used while searching, as shown in Table 9.1, and the criteria for exclusion

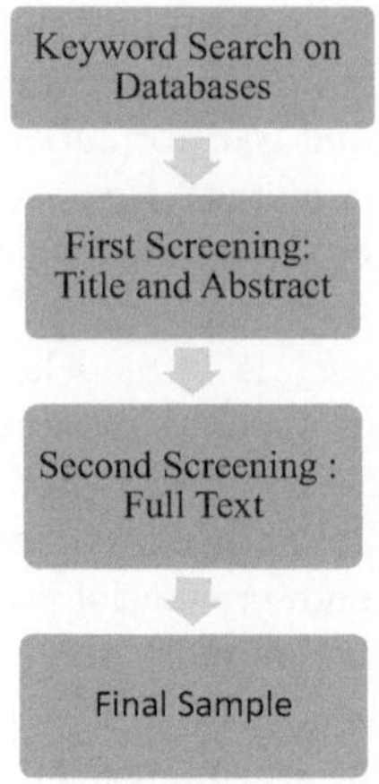

FIGURE 9.1 Screening process. (*Source:* Author.)

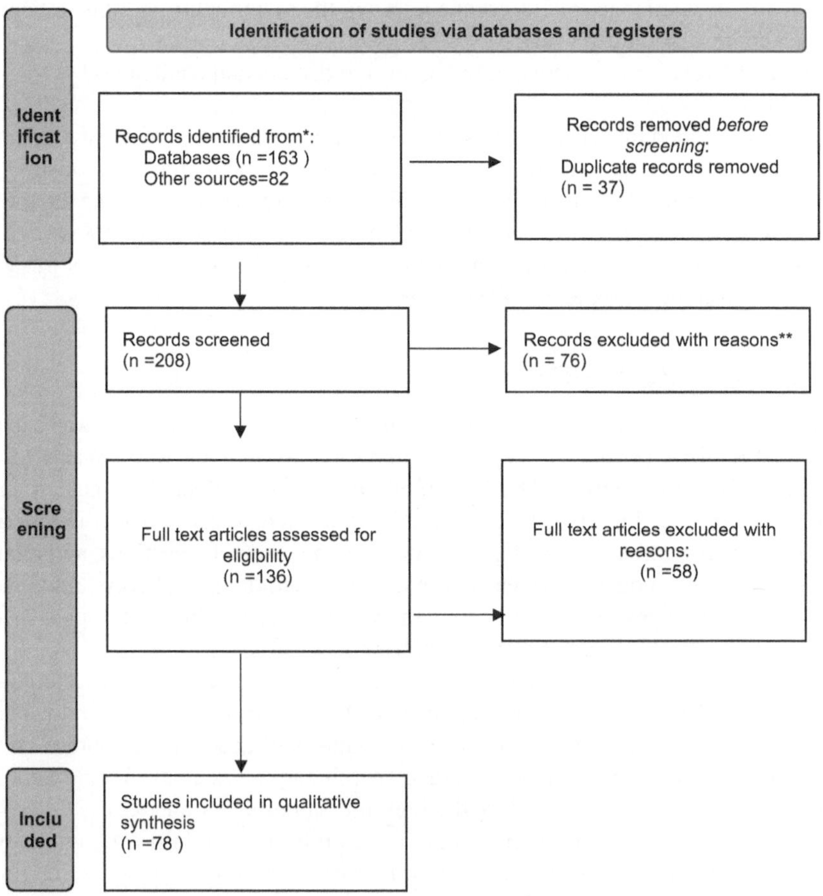

FIGURE 9.2 PRISMA FLOW diagram. (*Source:* http://www.prisma-statement.org/ (Author's compilation).)

TABLE 9.1
Search Limitations and Keywords Used

Search limitations (databases and search engines)	Scopus, Web of science, springer Science Direct, Emerald, Industry Reports, and Google Scholar
Keywords	Fourth Industrial Revolution, Industry 4.0, Industry 4.0 Challenges, Entrepreneurship Challenges, Technological Drivers, Smart Manufacturing, Digitalization, Automation, Cyber Physical Systems

Source: Author.

TABLE 9.2
Criteria for Inclusion and Exclusion

Criteria	Inclusion	Exclusion
Type of literature	Journals (Indexed), chapters from books, proceeding of conferences, and Industry reports	Non-indexed journals, articles (magazine and newspaper)
Text	Full text	Not Full text
Language	English	Non-English
Timeline	2015–2021	Before 2015

Source: Author.

and inclusion are shown in Table 9.2. Four criteria were used for inclusion and exclusion, namely literature type, text, language, and timeline. As per the PRISMA approach, two rounds of elimination or screening were conducted, which led to the targeted 54 articles constituting this review and discussion.

A total of nine drivers and ten challenges have been identified after comprehensively reviewing the literature. Further, these identified drivers and challenges were discussed with a panel of experts to confirm their appropriateness. Further, the relationship between the selected drivers and challenges has been identified and the TISM methodology is used to prepare the structural model (Table 9.3).

9.2.1 Industry 4.0 in the Indian Context

The idea of I4.0 originated in Germany in 2011 as the German government's initiative to enhance the growth of the German manufacturing sector (Hizam-Hanafiah et al., 2020). The Fourth Industrial Revolution focuses on automation, digitalization, and replacement of old analog production systems with new automated data-driven tools (Zhou et al., 2016). The foundation of I4.0 is based on CPS, which has led to the integration of information and communication technologies, as well as helps in building an intelligent and digital factory (Karadayi-usta, 2019). Such factories bring a revolution in the manufacturing industry and have made the processes more digital, information-led, customized, and sustainable (Schneider, 2018). I4.0 uses

TABLE 9.3
Thematic Representation of Reviewed Papers

Themes	Sub Themes	Literature Source
Industry 4.0 (strategy and implementation)	I4.0 implementation, readiness models, I4.0 readiness in organizations, sustainability framework in i4.0, operational ambidexterity, development of local entrepreneurship in I4.0, approach of I4.0, industry leadership, continuous improvement strategies, holistic approach to Human Resource Management, regional innovation strategies, financial literacy of entrepreneurs, entrepreneurial competencies, entrepreneurial finance, strategic humanity in I4.0, hyper connected society, sustainable development in i4.0, future of manufacturing industries, conceptual model of i4.0, knowledge-based entrepreneurial ecosystems	Karadayi-usta, 2019, Hizam-Hanafiah et al., 2020, Sony and Naik, 2020, Kamble et al., 2018, Sahi et al., 2020, Kłobukowski and Pasieczny, 2020, Grenčíková et al., 2019, Mkwanazi and Mbohwa, 2018, Bonaccorsi et al., 2020, Vinodh et al., 2021, Chiarini et al., 2020a, Li, 2018, Miskiewicz and Kwilinski, 2020, Gentner, 2016, Ślusarczyk, 2018, Hecklau et al., 2016, Lepore and Spigarelli, 2020, Hatammimi and Krisnawati, 2018, Kruger and Steyn, 2020a, Vuong, 2020, Malik, 2019, Bauer et al., 2015, Batkovskiy et al., 2019, Formation of Global Competitive, 2020, Noerhartati et al., 2019, Ghobakhloo, 2018, Beltrami et al., 2021, Liao et al., 2017, Piccarozzi et al., 2018, Karacay, 2018, Entezari, 2015, Kandil, 2018
Technological drivers (in I4.0)	Operational drivers, technological development, digital transformation, entrepreneurial development, manufacturing innovation, AI & entrepreneurship, big data challenges, I4.0 technologies, maintenance transformation, application of i4.0 technologies in manufacturing, innovative solutions in logistics, innovation & entrepreneurship, business model innovation, drivers & barriers, digital trust, and human capital & AI	Stentoft et al., 2020, Chiarini et al., 2020b, Cantú-Ortiz et al., 2020, Kruger and Steyn, 2020b, Dean et al., 2021, Chalmers et al., 2020, Lee and Lim, 2021, M. Khan et al., 2017, da Silveira et al., 2019, Silvestri et al., 2020, Zheng et al., 2021, Witkowski, 2017, Ferreira and Lisboa, 2019, Müller, 2019, Stentoft et al., 2021, Mubarak and Petraite, 2020, Popkova and Sergi, 2020, Africa et al., 2020, Dalmarco et al., 2019
Challenges of entrepreneurship (in I4.0)	Challenges of sustainability, challenges of labor market in I4.0, challenges in production system in I4.0, managerial challenges, future industrial challenges, wireless requirement & challenges, youth entrepreneurship challenges, new talent management, supply chain sustainability, challenges of I4.0 technology, challenges of quality management, challenges in manufacturing SME's, cyber security challenges, and implementation challenges	Ghobakhloo, 2020, Kergroach, 2017, A. Khan and Turowski, 2016, A. Khan and Turowski, 2016, Schneider, 2018, Zhou et al., 2016, Varghese and Tandur, 2014, Müller et al., 2018, Bakator et al., 2018, Whysall et al., 2019, Luthra and Mangla, 2018, Prause, 2019, Prause, 2019, Loveder, 2017, Foidl and Felderer, 2016, Schröder, 2016, Culot et al., 2019, Bakhtari et al., 2020, Bondar, 2018, Bakhtari et al., 2020, Raeesi et al., 2016, Koleva, 2018, Abdul-Hamid et al., 2020, Mian et al., 2020

Source: Author.

technologies such as big data analytics, AI, the IoT, CPS, etc. to improve the productivity of manufacturing industrial systems and increase their efficiency. (Kruger and Steyn, 2020b)

Industry 4.0 is still a relatively new concept for developing countries, including India, and requires a thorough understanding and practice for its implementation and acceptance in business (Luthra and Mangla, 2018). The Indian manufacturing sector contributes 16%–17% to the Gross Domestic Product of the country and provides employment to about 12% of the workforce (IBEF_Annual_Report_2019–20.Pdf, n.d.). To further enhance the growth of the manufacturing sector, the Government of India launched an initiative called "Make in India" and several other projects, such as 'Digital India' and Smart City, to provide an opportunity for automation and technological advancements across the manufacturing sector (Luthra and Mangla, 2018).

In 2020, India ranked 88th out of 134 countries in Network Readiness Index (NRI). The score for the technological pillar was 76, which means there is rapid technological growth in the country that will increase the in coming years (Soumitra and Bruno, 2020). The adaptation levels of modern information technology have shown sluggish growth patterns in India as compared to other developing economies such as Malaysia and China, which have much lower NRIscores. The level of industrial automation in the manufacturing sector of India is also comparatively lower. From an entrepreneurial point of view, the implementation of I4.0-based concepts in the Indian manufacturing sector is in its emerging stage as compared to other sectors such as Fmcg, automobile, power, service management, and energy (Soumitra and Bruno, 2020).

Mechanization, electrification, and computerization are the three keywords of previous industrial revolutions. The keywords for the Fourth Industrial Revolution are automation and digitalization.

9.2.2 Technological Drivers of Industry 4.0

The concept of I4.0 is relatively new in a developing country like India. Table 9.4 contains a list of technological drivers or enablers responsible for the advent of I4.0 in India (Table 9.5).

TABLE 9.4
Technological Drivers of Industry 4.0

No.	Technological Driver	Description	Source
D1	Big data	Big data refers to the huge amount of data available and constantly growing databases. It uses advanced processing ways to extract valuable information swiftly from numerous data sources required for accurate decision-making. It allows the separation of important information from less important one.	Zhou et al., 2016, Khan et al., 2017, Mubarak and Petraite, 2020

(Continued)

TABLE 9.4 (*Continued*)
Technological Drivers of Industry 4.0

No.	Technological Driver	Description	Source
D2	CPS	CPS is an interconnection of a cyber-twin and a physical component, where a computer program is a representative of a cyber-twin in the form of a simulation model. Different CPS are connected through IoT, for information transfer. This means that IoT acts as a connection bridge between various CPS.	Karacay, 2018, Karadayi-usta, 2019, K et al., 2020, Stentoft et al., 2021
D3	IoT	IoT, also referred to as internet of everything, is a structure in which data are exchanged between the physical world and computers via sensors, laser scanners, and other similar devices. Three features that characterize IoT are optimization, context, and omnipresence. IoT plays a very crucial role in I4.0,	Zhou et al., 2016, Lampropoulos et al., 2019, Cantú-Ortiz et al., 2020, Varghese and Tandur, 2014
D4	Operational efficiency	A technologically updated production system comprising of I4.0 technologies like visual computing, CPS, AI, etc. It is more intelligent than any other traditional factory. However, technological upgradation requires both time and money for its implementation.	Zhou et al., 2016, Karacay, 2018, Prause, 2019, Witkowski, 2017, Sahi et al., 2020
D5	Industrial internet and cloud	It has emerged from a virtualization technology and involves multifaceted communication between products and its production process and all the machine data is organized into a cloud for future reference.	Zhou et al., 2016, Varghese and Tandur, 2014
D6	Digital supply chain	It is a customer-driven, agile platform model that uses real-time data from various sources and analyzes it to predict demand, stimulation, and matching to optimize performance and eliminate risk.	Agrawal et al., 2020, Zheng et al., 2021
D7	Artificial intelligence	It is the capability of a system to understand external data properly and learn from this data. This data further use algorithms to accurately predict a phenomenon. Such data sets and algorithms can further be turned toward an entrepreneurial opportunity identification and exploitation.	Cantú-Ortiz et al., 2020, Chalmers et al., 2020, Lee and Lim, 2021
D8	Advanced manufacturing solutions	Rethinking old business models by implementing smarter ways of operation using high-tech machinery and digital devices. Software platforms enabled by internet and cloud computing have made the exchange of data simpler and has increased the interconnectedness among various levels in a company.	Cantú-Ortiz et al., 2020, Chalmers et al., 2020, Karacay, 2018
D9	Augmented reality	Augmented reality-based system can support different services, such as choosing parts in a warehouse and sending out orders over mobile devices	Chalmers et al., 2020, Witkowski, 2017

Source: Author's own.

TABLE 9.5
Challenges of Entrepreneurship in Industry 4.0

No.		Description	Source
C1	High initial cost of infrastructure	One of the most significant barriers to entrepreneurship in I4.0 is high investment cost associated with the installation of new digital technologies, resources, and skilled workforce for making their efforts profitable.	Agrawal et al., 2020, Bakhtari et al., 2020
C2	Lack of education and skills training program	The education system in India is not skill-based, especially in terms of technological operations, thus the workforce lacks the skill and training required for operation of smart factories. Therefore, manufacturing industries need to develop training program for education and skill enhancement so that their employees can align their skills with the newest technologies being used in I4.0.	Bakhtari et al., 2020, Cantú-Ortiz et al., 2020, Mian et al., 2020, Karacay, 2018
C3	Lack of technological integration and compatibility	The integration of technological tools (e.g. AI, IoT) with the infrastructure of manufacturing industries is essential for establishment of smart factories and is the major requirement for their compatibility with advance tools and techniques.	Raeesi et al., 2016, Bakhtari et al., 2020)
C4	Lack of skilled workforce	Lack of skilled workforce poses as a hindrance in the implementation of I4.0. Advance production processes require workers with new set of skills and competencies required to overcome challenges that arise due to digital transformation. Without a technologically friendly workforce, the benefits associated with digitization cannot be brought to use.	Cantú-Ortiz et al., 2020, Agrawal et al., 2020
C5	Lack of sustainable business models	Sustainable business models contribute to environmental and social sustainability through efficient resource usage and reduction in waste. The integration of I4.0 and sustainability is defined through concepts like green cloud computing, sustainable supply chains, etc.	Bakhtari et al., 2020, Beltrami et al., 2021, Batkovskiy et al., 2019
C6	Aversion to risk	It is defined as lack of willingness to undertake risk. This is a psychological barrier or challenge for entrepreneurship and has a negative effect on it. Tolerance of risk is a desired attitude for entrepreneurs.	Raeesi et al., 2016, Ghobakhloo, 2018, Kruger and Steyn, 2020a
C7	Lack of entrepreneurial capacity	Entrepreneurial capacity is the degree to which an individual is motivated to realize entrepreneurial opportunities and has the capacity, skill, and willingness to exploit entrepreneurial activities.	Raeesi et al., 2016, Kruger and Steyn, 2020a
C8	Organizational constraint	Traditional business processes do not fit in digital business models and must be optimized to meet the requirements of global value chains. Flexibility is essential to overcome the challenge of increasing product portfolio and customer demands.	Raeesi et al., 2016, Kruger and Steyn, 2020a, Zheng et al., 2021

(Continued)

TABLE 9.5 (*Continued*)
Challenges of Entrepreneurship in Industry 4.0

No.		Description	Source
C9	Lack of standardization	The emergence of Industry 4.0 has enabled inter-company integration and networking; thus standardization is required for information exchange and integration processes of cyber physical systems with traditional business models.	Raeesi et al., 2016, Bakhtari et al., 2020
C10	Digital security (digital trust)	With the application of technology, there will be immense interconnectedness and data flow; these data may include confidential information related to organization, its customers, and products. Thus, the companies need to ensure that these data and information are well-protected from damage, unauthorized accesses, and hackings. Security is a critical issue as digitalization and distribution of company information with third parties increase data exposure risk.	Agrawal et al., 2020, Bakhtari et al., 2020

Source: Author's own.

9.3 RESEARCH METHODOLOGY

9.3.1 The Method

TISM is an upgraded qualitative modeling technique of the traditional interpretive structure modeling (ISM). ISM was initially developed by J. Warfield in 1973. TISM is ascertained as a decision-modeling approach and has been used by numerous researchers (Sushil, 2013; Yadav, 2014; Dubey et al., 2017).

The process of TISM is presented in the following steps:

1. The primary step is to identify and define the variables or factors whose relations are to be modeled. In this research paper, the identification of factors is done by reviewing the existing literature.
2. Next, contextual relations are determined by examining the relationships among the identified variables.
3. This is the step toward the TISM that is required to further interpret the traditional ISM. Here, how one variable will influence or enhance the other variable is mentioned. For pair-wise comparisons, each factor is individually linked with all the other factors in the form of a reachability matrix. The total number of pair-wise comparisons for 'n' identified variables will be $n \times (n-1)$. The paired comparisons in the interpretive logic-knowledge base are described in the reachability matrix by making entry 1 in i–j cell if the corresponding entry in knowledge is "*Y*" i.e. Yes, which is further interpreted and by making an entry as 0 for the corresponding entry "*N*" i.e. No in the knowledge base.

4. Then, a transitivity check is done on this reachability matrix. The knowledge base is updated as "*Y*" and 1 is entered for each new transitive link and transitive is entered in the interpretation column. If the transitive link is significant in nature, its reason is also stated.
5. The next step is to ascertain and establish the reachability and antecedent sets for all the identified variables. The intersection of the reachability and the antecedent set will be the same as the reachability set in the case of the variable at a particular level. The top-level variable satisfying the condition should be removed from the variable set, and this will be repeated to determine all the levels.
6. Finally, the structural model is developed by arranging the variables graphically level-wise and the directed links are drawn as per the relations in the reachability matrix. Only those transitive relationships are retained in the digraph whose interpretation is crucial (Jena et al., 2017).

9.3.2 Sample and Data Collection

Data were collected from both secondary and primary sources. Existing literature on entrepreneurship, Industry 4.0, and related topics was reviewed to identify the variables or factors.

For the qualitative modeling, primary data has been collected from two industry experts. One is collected from academia through two structured templates: one each for technological drivers and challenges for drawing relationships between the identified variables. Their responses were sought as "yes" or "no", along with interpretation for determining relationships among factors and the majority view opinion i.e. two out of three were taken in the making of TISM.

9.4 RESULTS, ANALYSIS, AND DISCUSSION

9.4.1 Technological Drivers

The responses of the experts obtained through the template are arranged in the reachability matrix (Tables 9.6–9.10).

As per Figure 9.3, the nine identified drivers have been apportioned into two levels. Big data, IoT, industrial internet and cloud, AI, and augmented reality are on Level 2 and CPS, operational efficiency, digital supply chain, and advanced manufacturing solutions are on Level 1. Level 2 factors are driving factors, whereas Level 1 factors are emergent factors.

Big data consists of a large amount of data sets and information. Big data analytics is used to extract valuable information, bring about a logical understanding of it, and help in making informed decisions by various stakeholders. The flow of information of the data is done through IoT. IoT are physical objects such as laptops, mobiles, smart security cameras, etc. that provide internet connectivity through different software applications and sensors built into them. They can be used as controllers for other gadgets and appliances. IoT also enables storage and access of data on the cloud. Cloud facilitates the storage of data on servers via the internet by acting

TABLE 9.6
Initial Reachability Matrix

	D1	D2	D3	D4	D5	D6	D7	D8	D9
D1	1	1	0	0	1	1	1	0	1
D2	0	1	1	0	0	1	0	1	0
D3	1	1	1	1	1	1	1	1	1
D4	0	1	0	1	0	1	0	0	0
D5	1	1	1	1	1	1	1	1	1
D6	0	0	0	1	0	1	0	1	0
D7	1	1	1	1	1	1	1	1	1
D8	0	1	0	1	0	1	0	1	0
D9	0	1	1	1	0	1	0	1	1

TABLE 9.7
Final Reachability Matrix

	D1	D2	D3	D4	D5	D6	D7	D8	D9
D1	1	1	1[a]	1[a]	1	1	1	1[a]	1
D2	1[a]	1	1	1[a]	1[a]	1	1[a]	1	1[a]
D3	1	1	1	1	1	1	1	1	1
D4	0	1	1[a]	1	0	1	0	1[a]	0
D5	1	1	1	1	1	1	1	1	1
D6	0	1[a]	0	1	0	1	0	1	0
D7	1	1	1	1	1	1	1	1	1
D8	1[a]	1	1[a]	1	0	1	0	1	0
D9	1[a]	1	1	1	1[a]	1	1[a]	1	1

Source: Author own analysis.
[a] Value after applying transitivity.

as a database and allows remote access to the data through IoT. Data on the cloud is used by AI models for analysis and predicting future outcomes to aid organizations in taking strategic decisions. AI also helps in the automation of various processes. It is also adding value to the existing system of cloud computing, which has led to an integration of AI and the cloud. This has led to the formation of augmented reality. Augmented reality is an extension of virtual reality. It brings about the virtual objects in the real environment and provides a reality-based experience.

Big data also provides input to CPS. CPS encompasses physical components and digital components and is interconnected. CPS models help in monitoring, controlling, and automation of processes. It is also known as intelligent systems. These systems use IoT, cloud, AI, and augmented reality for the successful running their models. In terms of the manufacturing sector in Industry 4.0, digital transformation

TABLE 9.8
Iteration-1

Factors	Reachability Set	Antecedent Set	Intersection Set	Level
D1	1, 2, 3, 4, 5, 6, 7, 8, 9	1, 2, 3, 5, 7, 8, 9	1, 2, 3, 5, 7, 8, 9	
D2	**1, 2, 3, 4, 5, 6, 7, 8, 9**	1, 2, 3, 4, 5, 6, 7, 8, 9	**1, 2, 3, 4, 5, 6, 7, 8, 9**	I
D3	1, 2, 3, 4, 5, 6, 7, 8, 9	1, 2, 3, 4, 5, 7, 8, 9	1, 2, 3, 4, 5, 7, 8, 9	
D4	**2, 3, 4, 6, 8**	1, 2, 3, 4, 5, 6, 7, 8, 9	**2, 3, 4, 6, 8**	I
D5	1, 2, 3, 4, 5, 6, 7, 8, 9	1, 2, 3, 5, 7, 9	1, 2, 3, 5, 7, 9	
D6	**2, 4, 6, 8**	1, 2, 3, 4, 5, 6, 7, 8, 9	**2, 4, 6, 8**	I
D7	1, 2, 3, 4, 5, 6, 7, 8, 9	1, 2, 3, 5, 7, 9	1, 2, 3, 5, 7, 9	
D8	**1, 2, 3, 4, 6, 8**	1, 2, 3, 4, 5, 6, 7, 8, 9	**1, 2, 3, 4, 6, 8**	I
D9	1, 2, 3, 4, 5, 6, 7, 8, 9	1, 2, 3, 5, 7, 9	1, 2, 3, 5, 7, 9	

Source: Author own analysis.

TABLE 9.9
Iteration-2

Factors	Reachability Set	Antecedent Set	Intersection Set	Level
D1	**1, 3, 5, 7, 9**	1, 3, 5, 7, 9	**1, 3, 5, 7, 9**	II
D3	**1, 3, 5, 7, 9**	1, 3, 5, 7, 9	**1, 3, 5, 7, 9**	II
D5	**1, 3, 5, 7, 9**	1, 3, 5, 7, 9	**1, 3, 5, 7, 9**	II
D7	**1, 3, 5, 7, 9**	1, 3, 5, 7, 9	**1, 3, 5, 7, 9**	II
D9	**1, 3, 5, 7, 9**	1, 3, 5, 7, 9	**1, 3, 5, 7, 9**	II

Source: Author own analysis.

TABLE 9.10
List of Variables and Their Levels in TISM

S. No.	Variable No.	Variables	Level in TISM
1.	D1	Big Data	II
2.	D2	Cyber Physical Systems (CPS)	I
3.	D3	Internet of Things (IoT)	II
4.	D4	Operational Efficiency (OE)	I
5.	D5	Industrial Internet and Cloud	II
6.	D6	Digital Supply Chain (DSC)	I
7.	D7	Artificial Intelligence (AI)	II
8.	D8	Advanced Manufacturing Solutions	I
9.	D9	Augmented Reality	II

Source: Author own analysis.

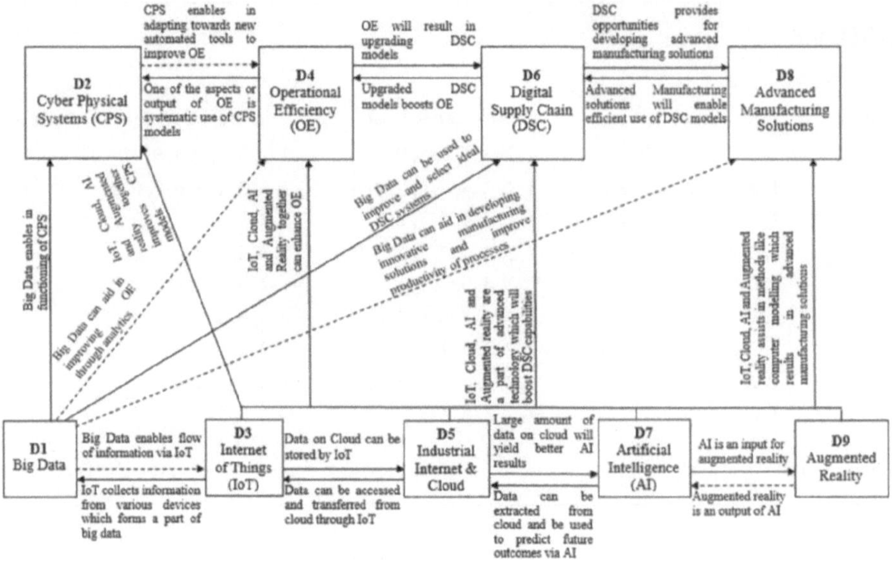

FIGURE 9.3 Hierarchical TISM of technological drivers in Industry 4.0.

has aided in operational efficiency by making the industrial production processes cost-effective and improving the quality of the output of each process. Digital transformation has also assisted in better inventory control of raw materials, quality control of products & processes, and reduction of defects of materials that have positively impacted operational efficiency. Big data has facilitation of upgradation of the existing supply chain systems to digital supply chain by enlarging the existing data sets for analysis and enabling the application of its analytic tools in selecting ideal digital supply chain models. All of these advancements have led to the development of advanced manufacturing solutions, such as flexible plant & production lines, optimum utilization of available resources, and intelligent manufacturing support.

In 2018, the World Economic Forum established its hub to promote the fourth Industrial Revolution, in India, in collaboration with the Government of India. The National Institute for Transforming India (NITI) Aayog has been assigned to work along with the World Economic Forum for expanding the new policy frameworks for upcoming digital technologies. (Kulshreshtha, 2021)

The Indian Institute of Science, India, launched a smart factory platform for manufacturing in 2014. This platform aims to design an industrial environment of intelligent technology and support automation intensive factory-like 3D printers, automated-guided vehicles, etc. (*Overview – Industry 4.0 : I4.0@IISc*, n.d.)

9.4.2 Challenges

The initial reachability matrix of challenges (Tables 9.11 and 9.12) was created using the same methods as technological drivers.

TABLE 9.11
Initial Reachability Matrix

	C1	C2	C3	C4	C5	C6	C7	C8	C9	C10
C1	1	0	1	0	1	0	0	0	1	0
C2	0	1	0	1	1	0	1	1	0	0
C3	0	0	1	0	1	0	0	1	1	0
C4	0	0	1	1	1	0	0	1	0	0
C5	0	0	1	0	1	0	0	0	0	0
C6	0	0	0	0	0	1	0	1	1	0
C7	0	0	0	0	1	1	1	1	1	0
C8	1	0	0	0	1	0	0	1	1	0
C9	0	0	1	0	0	0	0	0	1	0
C10	0	0	0	0	0	0	0	0	0	1

Source: Author own analysis.

TABLE 9.12
Final Reachability Matrix

	C1	C2	C3	C4	C5	C6	C7	C8	C9	C10
C1	1	0	1	0	1	0	0	1[a]	1	0
C2	1[a]	1	1[a]	1	1	1[a]	1	1	1[a]	0
C3	1[a]	0	1	0	1	0	0	1	1	0
C4	1[a]	0	1	1	1	0	0	1	1[a]	0
C5	0	0	1	0	1	0	0	1[a]	1[a]	0
C6	1[a]	0	1[a]	0	1[a]	1	0	1	1	0
C7	1[a]	0	1[a]	0	1	1	1	1	1	0
C8	1	0	1[a]	0	1	0	0	1	1	0
C9	0	0	1	0	1[a]	0	0	1[a]	1	0
C10	0	0	0	0	0	0	0	0	0	1

Source: Author own analysis.
[a] Value after applying transitivity.

9.4.2.1 Partitioning of the Final Reachability Matrix into Different Levels

As per Figure 9.4, the 10 identified challenges have been apportioned into five levels. Lack of education and skills training program is on level 5 and is the most independent and driving factor. Lack of entrepreneurial capacity is on level 4. Level 3 consists of a lack of skilled workforce (Workers 4.0) and aversion to risk. The high initial investment cost of infrastructure is on Level 2. Lack of technological integration and compatibility, lack of sustainable business models, organizational constraints, lack of standardization, and digital security (digital trust) are consequences of the higher levels and therefore they are placed on Level 1 (Tables 9.13–9.17).

TABLE 9.13
Iteration-1

Factors	Reachability Set	Antecedent Set	Intersection Set	Level
C1	1, 3, 5, 8, 9	1, 2, 3, 4, 6, 7, 8	1, 3, 8	
C2	1, 2, 3, 4, 5, 6, 7, 8, 9	2	2	
C3	**1, 3, 5, 8, 9**	1, 2, 3, 4, 5, 6, 7, 8, 9	**1, 3, 5, 8, 9**	I
C4	1, 3, 4, 5, 8, 9	2, 4	4	
C5	**3, 5, 8, 9**	1, 2, 3, 4, 5, 6, 7, 8, 9	**3, 5, 8, 9**	I
C6	1, 3, 5, 6, 8, 9	2, 6, 7	6	
C7	1, 3, 5, 6, 7, 8, 9	2, 6, 7	6, 7	
C8	**1, 3, 5, 8, 9**	1, 2, 3, 4, 5, 6, 7, 8, 9	**1, 3, 5, 8, 9**	I
C9	**3, 5, 8, 9**	1, 2, 3, 4, 5, 6, 7, 8, 9	**3, 5, 8, 9**	I
C10	**10**	10	**10**	I

Source: Author own analysis.

TABLE 9.14
Iteration-2

Factors	Reachability Set	Antecedent Set	Intersection Set	Level
C1	1	1, 2, 4, 6, 7	1	II
C2	1, 2, 4, 6, 7	2	2	
C4	1, 4	2, 4	4	
C6	1, 6	2, 6, 7	6	
C7	1, 6, 7	2, 6, 7	6, 7	

Source: Author own analysis.

TABLE 9.15
Iteration-3

Factors	Reachability Set	Antecedent Set	Intersection Set	Level
C2	2, 4, 6, 7	2	2	
C4	**4**	2, 4	**4**	III
C6	**6**	2, 6, 7	**6**	III
C7	6, 7	2, 6, 7	6, 7	

Source: Author own analysis.

TABLE 9.16
Iteration-4

Factors	Reachability Set	Antecedent Set	Intersection Set	Level
C2	2, 7	2	2	V
C7	**7**	2, 7	7	IV

Source: Author own analysis.

TABLE 9.17
List of Variables and Their Levels in TISM

S. No.	Variable No.	Variables	Level in TISM
1.	C1	High initial investment cost on infrastructure	II
2.	C2	Lack of education and skills training program	V
3.	C3	Lack of technological integration and compatibility	I
4.	C4	Lack of skilled workforce (Worker 4.0)	III
5.	C5	Lack of sustainable business models	I
6.	C6	Aversion to risk	III
7.	C7	Lack of entrepreneurial capacity	IV
8.	C8	organizational constraints	I
9.	C9	Lack of standardization	I
10.	C10	Data security (digital trust)	I

Source: Author own analysis.

Education and skills training program in the context of Industry 4.0 is the backbone of successful entrepreneurship ventures. The era of Industry 4.0 is new and still evolving. It becomes very important for entrepreneurs to acquire knowledge about new developments, especially digital transformations, and receive hands-on training of these applications. Lack of these will create shortage of skilled workers in this industry and limit entrepreneurial competence.

In India, an important initiative has been undertaken by the Department of Heavy Industries (Ministry of Heavy Industries and Public Enterprises), Government of India, known as "SAMARTH Udyog Bharat 4.0" to facilitate 4.0 manufacturing projects. It attempts to overcome the challenge of lack of education and training programs (*About I4.0 | SAMARTH Udyog*, n.d.).

The lack of entrepreneurial capacity will give rise to excessive aversion of risk behavior by the entrepreneurs. This can result in the wrong decision-making, leading to high initial investment cost of infrastructure. This in turn will become a hindrance to technological integration and compatibility, which will cause inefficiency and wastage of resources. This wastage will work against the development of sustainable business models. Organizational constraints like traditional business models and inflexible business processes can also hamper the development of sustainable business models. It also leads to lack of standardization. Standardization refers to the

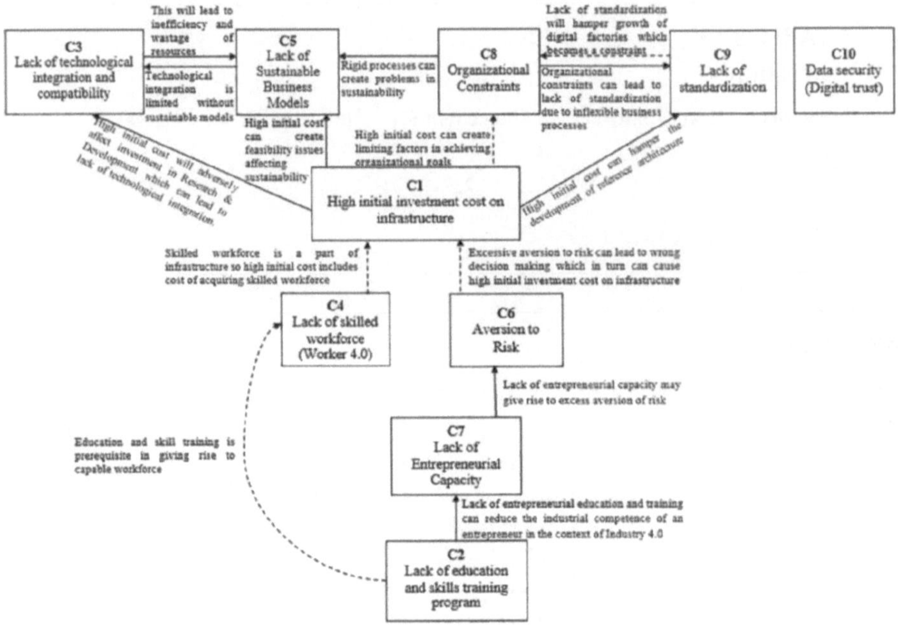

FIGURE 9.4 Hierarchical TISM of challenges in Industry 4.0.

systematic process of information exchange and adaptable technology that can be used by all departments of an organization.

The factor of digital security (digital trust) does not have any relationship with the identified challenges. However, it is important to recognize the cyber security issues, loopholes in the system, unauthorized access & use of data, and cyber-attacks so that it may instill trust among people to shift their business operations and the relevant processes to online platforms.

9.5 CONCLUSION

The study presents a comprehensive analysis by applying TISM to the enablers of Industry 4.0, how they are interlinked, and the critical driving technologies that are triggering the implementation of Industry 4.0 within the manufacturing industry of India. The research study enhances the basic understanding of the Industry 4.0 concept and the technologies driving it. The full interpretive structural modeling exhibits how technologies such as big data, IoT, AI, and augmented reality considerably derive the inception of CBS, which are now essential for efficient operations in the manufacturing industry. Operational efficiency, digital supply chain, and advanced manufacturing solutions are only possible when the basic framework of driving technological enablers is appropriately executed.

However, after analyzing the challenges to entrepreneurship via TISM, it became clear how different entrepreneurial challenges hinder the acceptance of Industry 4.0 and its technologies in developing countries like India. Therefore, the research study vividly describes how these entrepreneurial barriers restrict the implementation of the significant technologies central to Industry 4.0. There are several government initiatives up and running in India to equip the workforce with the required skills and expertise essential within Industry 4.0. However, there must be more such initiatives demanded by the manufacturing industry to operate its best and contribute significantly to the country's gross domestic product. Also, by focusing on education and skills training program, the other challenges can be overcome as the most independent factor is 'lack of education and skills training program'. Along with this, the issue of cyber security has to be dealt with stringently to encourage businesses to align their operations in accordance with Industry 4.0.

9.6 FUTURE RESEARCH DIRECTIONS

This study suggests nine technological drivers of Industry 4.0 and 10 challenges of entrepreneurship in context of the manufacturing industry of India. For further research, some of the other challenges in another country or context or non-technological drivers may be included. This work has been conducted in the Indian context. This can be extended to other developing nations as well. Furthermore, the drivers and challenges of entrepreneurship in different developing nations can be compared in future studies. Also, more responses can be collected to make the TISM more accurate. There is scope for doing similar types of research with ISM or TISM technique with an emphasis on other developing countries.

REFERENCES

Abdul-Hamid, A. Q., Ali, M. H., Tseng, M. L., Lan, S., and Kumar, M. (2020). Impeding challenges on industry 4.0 in circular economy: Palm oil industry in Malaysia. *Computers and Operations Research*, 123, 105052. https://doi.org/10.1016/j.cor.2020.105052

About I4.0 | SAMARTH Udyog. (n.d.).

Africa, S., Faculty, B. M., Sciences, M., and Africa, S. (2020). Predictors of industry 4.0 technologies affecting logistic enterprises' performance: International perspective from economic lens. *Technological and Economic Development of Economy*, 26(6), 1263–1283.

Agrawal, P., Narain, R., and Ullah, I. (2020). Analysis of barriers in implementation of digital transformation of supply chain using interpretive structural modelling approach. *Journal of Modelling in Management*, 15(1), 297–317. https://doi.org/10.1108/JM2-03-2019-0066.

Bakator, M., Đorđević, D., Ćoćkalo, D., Nikolić, M., and Vorkapić, M. (2018). Lean startups with industry 4.0 technologies: Overcoming the challenges of youth entrepreneurship in Serbia. *Journal of Engineering Management and Competitiveness*, 8(2), 89–101. https://doi.org/10.5937/jemc1802089b.

Bakhtari, A. R., Kumar, V., Waris, M. M., Sanin, C., and Szczerbicki, E. (2020). Industry 4. 0 implementation challenges in manufacturing industry implementation challenges in manufacturing industries: An interpretive structural modelling approach industries: An interpretive structural modelling approach. *Procedia Computer Science*, 176, 2384–2393. https://doi.org/10.1016/j.procs.2020.09.306.

Batkovskiy, A. M., Leonov, A. V, Pronin, A. Y., Semenova, E. G., Fomina, A. V, and Balashov, V. M. (2019). Sustainable development of industry 4.0: The case of high-tech products system design. *Entrepreneurship and Sustainability Issues*, 6(4), 1823–1838.

Bauer, W., Hämmerle, M., Schlund, S., and Vocke, C. (2015). Transforming to a hyper-connected society and economy – towards an " Industry 4. 0." *Procedia Manufacturing*, 3(Ahfe), 417–424. https://doi.org/10.1016/j.promfg.2015.07.200

Beltrami, M., Orzes, G., Sarkis, J., and Sartor, M. (2021). Industry 4.0 and sustainability: Towards conceptualization and theory. *Journal of Cleaner Production*, 312, 127733. https://doi.org/10.1016/j.jclepro.2021.127733

Bonaccorsi, A., Chiarello, F., Fantoni, G., and Kammering, H. (2020). Emerging technologies and industrial leadership.A Wikipedia-based strategic analysis of Industry 4.0. *Expert Systems With Applications*, 113645. https://doi.org/10.1016/j.eswa.2020.113645.

Bondar, K. (2018). Challenges and Opportunities of Industry 4.0 – Spanish Experience. *International Journal of Innovation, Management and Technology*, 9(5), 202–208. https://doi.org/10.18178/ijimt.2018.9.5.814.

Cantú-Ortiz, F. J., Galeano Sánchez, N., Garrido, L., Terashima-Marin, H., and Brena, R. F. (2020). An artificial intelligence educational strategy for the digital transformation. *International Journal on Interactive Design and Manufacturing*, 14(4), 1195–1209. https://doi.org/10.1007/s12008-020-00702-8.

Chalmers, D., MacKenzie, N. G., and Carter, S. (2020). Artificial intelligence and entrepreneurship: Implications for venture creation in the fourth industrial revolution. *Entrepreneurship:TheoryandPractice*,1–26.https://doi.org/10.1177/1042258720934581

Chiarini, A., Belvedere, V., and Grando, A. (2020a). Industry 4.0 strategies and technological developments. An exploratory research from Italian manufacturing companies. *Production Planning and Control*, 31(16), 1385–1398. https://doi.org/10.1080/09537287.2019.1710304

Chiarini, A., Belvedere, V., and Grando, A. (2020b). The management of operations industry 4. 0 strategies and technological developments. An exploratory research from Italian manufacturing companies. *Production Planning and Control*, 0(0), 1–14. https://doi.org/10.1080/09537287.2019.1710304.

Culot, G., Fattori, F., Podrecca, M., and Sartor, M. (2019). Addressing industry 4.0 cybersecurity challenges. *IEEE Engineering Management Review*, 47(3), 79–86. https://doi.org/10.1109/EMR.2019.2927559.

da Silveira, F., Neto, I. R., Machado, F. M., da Silva, M. P., and Amaral, F. G. (2019). Analysis of industry 4.0 technologies applied to the health sector: Systematic literature review. *Studies in Systems, Decision and Control*, 202, 701–709. https://doi.org/10.1007/978-3-030-14730-3_73.

Dalmarco, G., Ramalho, F. R., Barros, A. C., and Soares, A. L. (2019). Providing industry 4.0 technologies: The case of a production technology cluster. *Journal of High Technology Management Research*, 30(2), 100355. https://doi.org/10.1016/j.hitech.2019.100355.

Dean, M., Rainnie, A., Stanford, J., and Nahum, D. (2021). Industrial policy-making after COVID-19: Manufacturing, innovation and sustainability. *Economic and Labour Relations Review*, 32(2), 283–303. https://doi.org/10.1177/10353046211014755.

Entezari, Y. (2015). Building knowledge – based entrepreneurship ecosystems: Case of. *Procedia – Social and Behavioral Sciences*, 195, 1206–1215. https://doi.org/10.1016/j.sbspro.2015.06.242.

Ferreira, V., and Lisboa, A. (2019). Innovation and entrepreneurship: From schumpeter to industry 4.0. *Applied Mechanics and Materials* 890, 174–180. https://doi.org/10.4028/www.scientific.net/AMM.890.174.

Foidl, H., and Felderer, M. (2016). Research challenges of industry 4.0 for quality management. *Lecture Notes in Business Information Processing*, 245, 121–137. https://doi.org/10.1007/978-3-319-32799-0_10

Dzwigol, H., Dzwigol-Barosz, M., Kwilinski, A. (2020). Formation of global competitive enterprise environment based on industry 4.0 concept. *International Journal of Entrepreneurship*, 24(1), 4675.

Gentner, S. (2016). Industry 4. 0: Reality, future or just science fiction? How to convince today's management to invest in tomorrow's future! Successful strategies for industry 4.0 and manufacturing IT. *Chimia* 70(9), 628–633. https://doi.org/10.2533/chimia.2016.628.

Ghobakhloo, M. (2018). The future of manufacturing industry: A strategic roadmap toward Industry 4.0. *Journal of Manufacturing Technology Management*, 29(6), 910–936. https://doi.org/10.1108/JMTM-02-2018-0057

Ghobakhloo, M. (2020). Industry 4.0, digitization, and opportunities for sustainability. *Journal of Cleaner Production*, 252, 119869. https://doi.org/10.1016/j.jclepro.2019.119869.

Grenčíková, A., Kordoš, M., and Sokol, J. (2019). The approach to industry 4.0 within the Slovak business environment. *Social Sciences*, 8(4). https://doi.org/10.3390/socsci8040104.

Hatammimi, J., and Krisnawati, A. (2018). Financial literacy for entrepreneur in the industry 4.0 era: A conceptual framework in Indonesia. *ACM International Conference Proceeding Series*, 183–187. https://doi.org/10.1145/3285957.3285985.

Hecklau, F., Galeitzke, M., Flachs, S., and Kohl, H. (2016). Holistic approach for human resource management in industry 4.0. *Procedia CIRP*, 54, 1–6. https://doi.org/10.1016/j.procir.2016.05.102

Hizam-Hanafiah, M., Soomro, M. A., and Abdullah, N. L. (2020). Industry 4.0 readiness models: A systematic literature review of model dimensions. *Information (Switzerland)*, 11(7), 1–13. https://doi.org/10.3390/info11070364

IBEF_Annual_Report_2019–20.pdf. (n.d.).

Jena, J., Sidharth, S., Thakur, L. S., Kumar Pathak, D., and Pandey, V. C. (2017). Total interpretive structural modeling (TISM): Approach and application. *Journal of Advances in Management Research*, 14(2), 162–181. https://doi.org/10.1108/JAMR-10-2016-0087.

Paranitharan, K. P., (2020). Interpretive framework by analyzing the enablers for implementation of Industry 4. 0: An ISM approach. *Total Quality Management and Business Excellence*, 1–21. https://doi.org/10.1080/14783363.2020.1735933.

Kamble, S. S., Gunasekaran, A., and Gawankar, S. A. (2018). Sustainable industry 4.0 framework: A systematic literature review identifying the current trends and future perspectives. *Process Safety and Environmental Protection*, 117, 408–425. https://doi.org/10.1016/j.psep.2018.05.009

Kandil, M. (2018). *The Eletronic Library*, 34(1), 1–5.

Karacay, G. (2018). *Talent Development for Industry 4.0.* https://doi.org/10.1007/978-3-319-57870-5_7.

Karadayi-usta, S. (2019). An Interpretive Structural Analysis for Industry 4.0. *IEEE Transactions on Engineering Management*, 1–6. https://doi.org/10.1109/TEM.2018.2890443

Kergroach, S. (2017). Industry 4.0: New challenges and opportunities for the labour market. *ForesightandSTIGovernance*, 11(4), 6–8. https://doi.org/10.17323/2500–2597.2017.4.6.8

Khan, A., and Turowski, K. (2016). A perspective on industry 4.0: From challenges to opportunities in production systems. *IoTBD 2016 – Proceedings of the International Conference on Internet of Things and Big Data, IoTBD*, 441–448. https://doi.org/10.5220/0005929704410448

Khan, M., Wu, X., Xu, X., and Dou, W. (2017). Big data challenges and opportunities in the hype of Industry 4.0. *IEEE International Conference on Communications*, 0–5. https://doi.org/10.1109/ICC.2017.7996801

Kłobukowski, P., and Pasieczny, J. (2020). Impact of resources on the development of local entrepreneurship in industry 4.0. *Sustainability*, 12(24), 10272.

Koleva, N. (2018). Industry 4.0's opportunities and challenges for production engineering and managemnet. *International Scientific Journals Innovations*, 6(1), 17–18.

Kruger, S., and Steyn, A. A. (2020a). A conceptual model of entrepreneurial competencies needed to utilize technologies of Industry 4. 0. *The International Journal of Entrepreneurship and Innovation*. 22(1), 56–67. https://doi.org/10.1177/1465750320927359.

Kruger, S., and Steyn, A. A. (2020b). Enhancing technology transfer through entrepreneurial development: Practices from innovation spaces. *Journal of Technology Transfer*, 45(6). https://doi.org/10.1007/s10961-019-09769-2.

Kulshreshtha, M. (2021). Industry 4.0 technology: The key game changer for Indian manufacturing sector (pp. 1–9).

Lampropoulos, G., Siakas, K., and Anastasiadis, T. (2019). Internet of things in the context of industry 4.0: An overview. *International Journal of Entrepreneurial Knowledge*, 7(1), 4–19. https://doi.org/10.2478/ijek-2019-0001.

Lee, C., and Lim, C. (2021). From technological development to social advance: A review of Industry 4.0 through machine learning. *Technological Forecasting and Social Change*, 167(January), 120653. https://doi.org/10.1016/j.techfore.2021.120653

Lepore, D., and Spigarelli, F. (2020). Integrating industry 4. 0 plans into regional innovation strategies. *Local Economy*. 35(5), 496–510. https://doi.org/10.1177/0269094220937452.

Li, L. (2018). China's manufacturing locus in 2025: With a comparison of "Made-in-China 2025" and "Industry 4.0." *Technological Forecasting and Social Change*, 135(May 2017), 66–74. https://doi.org/10.1016/j.techfore.2017.05.028

Liao, Y., Deschamps, F., de Loures, E. F. R., and Ramos, L. F. P. (2017). Past, present and future of Industry 4.0 - A systematic literature review and research agenda proposal. *International Journal of Production Research*, 55(12), 3609–3629. https://doi.org/10.1080/00207543.2017.1308576

Loveder, P. (2017). *Australian Apprenticeship: Trends, Challenges and Future Opportunities For Dealing with Industry 4.0.*

Luthra, S., and Mangla, S. K. (2018). Evaluating challenges to Industry 4.0 initiatives for supply chain sustainability in emerging economies. *Process Safety and Environmental Protection*. https://doi.org/10.1016/j.psep.2018.04.018

Malik, A. (2019). Creating competitive advantage through source basic capital strategic humanity in the industrial age 4.0. *International Research Journal of Advanced Engineering and Science*, 4(1), 209–215.

Mian, S. H., Salah, B., Ameen, W., Moiduddin, K., and Alkhalefah, H. (2020). Adapting universities for sustainability education in industry 4.0: Channel of challenges and opportunities. *Sustainability (Switzerland)*, 12(15). https://doi.org/10.3390/su12156100

Miskiewicz, R., and Kwilinski, A. (2020). *Entrepreneurship and Sustainability Issues* 7(4), 2630–2645.

Mkwanazi, S., and Mbohwa, C. (2018). Implications of the 4th industrial revolution on entrepreneurship education. *Proceedings of the International Conference on Industrial Engineering and Operations Management*, 2018(SEP), 318–328.

Mubarak, M. F., and Petraite, M. (2020). Technological forecasting and social change Industry 4.0 technologies, digital trust and technological orientation: What matters in open innovation ? *Technological Forecasting and Social Change*, 161(July), 120332. https://doi.org/10.1016/j.techfore.2020.120332.

Müller, J. M. (2019). Business model innovation in enterprises Strategies for industry 4.0 providers and users. *Journal of Manufacturing Technology Management* 30(8), 1127–1142. https://doi.org/10.1108/JMTM-01-2018-0008.

Müller, J. M., Kiel, D., and Voigt, K. I. (2018). What drives the implementation of Industry 4.0? The role of opportunities and challenges in the context of sustainability. *Sustainability (Switzerland)*, 10(1). https://doi.org/10.3390/su10010247

Noerhartati, E., Karyati, P. D., Soepriyono, S., and Yunarko, B. (2019). Entrepreneurship Sorghum towards Industry 4.0. *Proceedings of the International Conference on Innovation in Research (ICIIR 2018) – Section: Economics and Management Science*. https://doi.org/10.2991/iciir-18.2019.2

Overview – Industry 4.0 : I4.0@IISc. (n.d.).

Piccarozzi, M., Aquilani, B., and Gatti, C. (2018). Industry 4.0 in management studies: A systematic literature review. *Sustainability (Switzerland)*, 10(10), 1–24. https://doi.org/10.3390/su10103821

Popkova, E. G., and Sergi, B. S. (2020). Human capital and AI in industry 4.0. Convergence and divergence in social entrepreneurship in Russia. *Journal of Intellectual Capital*, 21(18), 565–581. https://doi.org/10.1108/JIC-09-2019-0224.

Prause, M. (2019). Challenges of industry 4. 0 technology adoption for SMEs: The case of Japan. *Sustainability*, 11(20), 5807.

Raeesi, R., Dastranj, M., Mohammadi, S., and Rasouli, E. (2016). Understanding the interactions among the barriers to entrepreneurship using interpretive structural modeling understanding the interactions among the barriers to entrepreneurship using interpretive structural modeling. *International Journal of Business and Management*, 8(13), 56–72. https://doi.org/10.5539/ijbm.v8n13p56.

Sahi, G. K., Gupta, M. C., and Cheng, T. C. E. (2020). The effects of strategic orientation on operational ambidexterity: A study of indian SMEs in the industry 4.0 era. *International Journal of Production Economics*, 220(August 2018), 107395. https://doi.org/10.1016/j.ijpe.2019.05.014.

Schneider, P. (2018). Managerial challenges of industry 4.0: An empirically backed research agenda for a nascent field. In *Review of Managerial Science* (Vol. 12, Issue 3). Springer, Berlin Heidelberg. https://doi.org/10.1007/s11846-018-0283-2.

Schröder, C. (2016). The challenges of industry 4.0 for small and medium-sized enterprises a good society – social democracy # 2017 plus. *The Friedrich-Ebert-Stiftung*, 28.

Silvestri, L., Forcina, A., Introna, V., Santolamazza, A., and Cesarotti, V. (2020). Maintenance transformation through industry 4.0 technologies: A systematic literature review. *Computers in Industry*, 123, 103335. https://doi.org/10.1016/j.compind.2020.103335.

Ślusarczyk, B. (2018). Industry 4.0 – Are we ready? *Polish Journal of Management Studies*, 17(1), 232–248. https://doi.org/10.17512/pjms.2018.17.1.19

Sony, M., and Naik, S. (2020). Key ingredients for evaluating Industry 4.0 readiness for organizations: A literature review. *Benchmarking*, 27(7), 2213–2232. https://doi.org/10.1108/BIJ-09-2018-0284.

Soumitra, D., and Bruno, L. (2020). *The Network Readiness Index 2020*, 2020(Figure 2), 1–6.

Stentoft, J., Adsbøll Wickstrøm, K., Philipsen, K., and Haug, A. (2021). Drivers and barriers for Industry 4.0 readiness and practice: Empirical evidence from small and medium-sized manufacturers. *Production Planning and Control*, 32(10), 811–828. https://doi.org/10.1080/09537287.2020.1768318

Stentoft, J., Wickstrøm, K. A., Philipsen, K., Haug, A., Stentoft, J., Wickstrøm, K. A., Philipsen, K., Stentoft, J., Wickstrøm, K. A., Philipsen, K., and Haug, A. (2020). The Management of Operations Drivers and barriers for Industry 4.0 readiness and practice: Empirical evidence from small and medium-sized manufacturers from small and medium-sized manufacturers. *Production Planning and Control*, 0(0), 1–18. https://doi.org/10.1080/09537287.2020.1768318

Varghese, A., and Tandur, D. (2014). Wireless requirements and challenges in Industry 4.0. *2014 International Conference on Contemporary Computing and Informatics (IC3I)*. 634–638.

Vinodh, S., Antony, J., Agrawal, R., and Douglas, J. A. (2021). Integration of continuous improvement strategies with Industry 4.0: A systematic review and agenda for further research. *TQM Journal*, 33(2), 441–472. https://doi.org/10.1108/TQM-07-2020-0157.

Vuong, Q. (2020). An unprecedented time for entrepreneurial finance upon the arrival of industry 4.0. *Journal of Risk and Financial Management*, 13(10), 224.

Whysall, Z., Owtram, M., and Brittain, S. (2019). The new talent management challenges of Industry 4.0. *Journal of Management Development*, 38(2), 118–129. https://doi.org/10.1108/JMD-06-2018-0181.

Witkowski, K. (2017). Internet of things, big data, industry 4.0 – Innovative solutions in logistics and supply chains management. *Procedia Engineering*, 182, 763–769. https://doi.org/10.1016/j.proeng.2017.03.197

Zheng, T., Ardolino, M., Bacchetti, A., and Perona, M. (2021). The applications of Industry 4.0 technologies in manufacturing context: A systematic literature review. *International Journal of Production Research*, 59(6), 1922–1954. https://doi.org/10.1080/00207543.2020.1824085

Zhou, K., Liu, T., and Zhou, L. (2016). Industry 4.0: Towards future industrial opportunities and challenges. *2015 12th International Conference on Fuzzy Systems and Knowledge Discovery, FSKD 2015*, 2147–2152. https://doi.org/10.1109/FSKD.2015.7382284

10 Toward Circular Economy and Social Equity

Role of Entrepreneurship

S. Senthil Kumar
Institute of Management Technology, Nagpur, India

CONTENTS

10.1 INTRODUCTION

The 'Brundtland Report' published in 1987 (World Commission on Environment and Development [WCED], 1987) marked the beginning of the discussion of SD in the academic domain. Development is considered sustainable if it upholds intergenerational equity, does not destroy environmental resources, is socially just, fair, and economically viable. Intergenerational equity can be established when the present generation's needs are met without affecting the future generation's ability to meet their own needs. SD recognizes the interdependence and interactions among social, economic, and ecological systems, and hence ecological sustenance, social equity, and economic stability are prioritized equally (Andriantiatsaholiniaina, Kouikoglou, & Phillis, 2004). SD acknowledges the finite capacity of natural stock while achieving economic growth, thus emphasising reduction or optimum-level material production and consumption (Hamstead & Quinn, 2005).

Entrepreneurship is widely celebrated for its role in the innovation-led growth and development of global economies. However, economic growth has also caused environmental degradation and social inequity. Hence, the concerned stakeholders expected entrepreneurship to provide solutions rather than contributing to unsustainable development. The recognition of entrepreneurship in addressing environmental degradation and social inequity stimulated research and practice of sustainability-driven entrepreneurship. From Schumpeter's view, the purpose of sustainability-oriented entrepreneurship can be described as innovating new products or processes or changes in existing products or processes, leading to the creative destruction of the unsustainable

DOI: 10.1201/9781003217084-10

products and processes used in the past (Schumpeter, 1942). Sustainability-driven entrepreneurship contributes to economic development while improving the quality of life of current and future generations and protecting natural capital.

The linear take-make-dispose production model is not sustainable in the long run due to supply and price risks associated with using virgin materials, the increasing trend of consumers preferring environmentally and socially responsible products, and stricter regulatory pressures (Moreno, Braithwaite, & Cooper, 2014). In contrast, the circular model offers potential opportunities for sustainability-oriented entrepreneurs to add more value by creating circulating assets for more extended and intensified product usage (Park, Sarkis, & Wu, 2010). The circular economy (CE) follows the closed-loop material flow, emphasizing the restorative or regenerative industrial system, where natural capital is used and reused as efficiently as possible. The three essential methods of closed-loop material flow are "reduce, reuse, and recycle" (3R). Reducing involves minimizing inputs in the value-adding process (supply) and consumer's consumption (demand). Reusing means using the waste obtained from a value-adding process as a raw material for another process. Recycling involves the transformation of used material into another product. McDonough and Braungart (2002) argued that 3R methods based on the 'cradle-to-grave' principle might not be good enough to achieve a circular and sustainable economy. They suggested two methods – downcycling and upcycling – that work on the radical principle of cradle-to-cradle. Downcycling means producing new lower-value products than the original ones through recycling methods. Upcycling involves recycling that adds new value to the recycled materials or items.

10.1.1 Opportunities for Sustainability-Oriented Entrepreneurship

Cohen and Winn (2007) proposed that market imperfections can be the source of opportunities for sustainable entrepreneurship. They identified four market imperfections – inefficient firms, externalities, imperfect pricing mechanisms, and information asymmetry – as the causal factors for environmental degradation. However, a comprehensive understanding of environmental challenges and market imperfections can lead to entrepreneurial innovations, which help mitigate environmental degradation or replenish natural resources.

Firms operating at sub-optimal production efficiencies waste an enormous amount of natural resources. Entrepreneurial opportunities can be identified in the methods to improve natural resource productivity, market functioning, and service delivery using 3R, downcycling, upcycling production principles of a CE. Negative externalities concerning the environment occur when the prices of products and services do not include the environmental cost incurred in the production process. Entrepreneurs can identify opportunities to generate positive environmental externalities through replacing old practices with new ones by leveraging technology and supply chain solutions. The prices of exhaustible natural resources used in the free market economy do not reflect their actual value, leading to skewed pricing. However, increased concerns for environmental protection led us to perfect pricing of natural resources, which stimulates demand for alternative, renewable factors of production such as wind, solar, geothermal, and recyclable materials. The search for renewable factors of production provides a host of entrepreneurial opportunities to create alternative technologies to harness renewable resources. The imperfect information on the environmental costs of production and consumption across the markets leads to an unsustainable economy. Entrepreneurs can develop products and services that provide information on the environmental impact of production and consumption practices, which eventually make economic agents and consumers opt for eco-efficient production methods, products, and services. For example, airveda – an entrepreneurial venture – produces a range of monitors that provide real-time information on air quality so that constant monitoring of air pollution levels in one's living area is possible.

Sustainability-driven entrepreneurship can be classified into eco-entrepreneurship, social entrepreneurship, and sustainable entrepreneurship, depending on the mission and the domains of value creation. This paper aims to analyze the entrepreneurial processes involved in achieving sustainability using case studies drawn from each domain.

10.2 RESEARCH METHODOLOGY

This study adopts the case study research method to draw inferences, suggestions for entrepreneurial practices, and necessary conclusions. The case study research method is used to analyze a contemporary phenomenon in its real-life context using data obtained from multiple sources (Yin, 1984). Case studies allow us to examine contemporary real-life phenomena through detailed contextual analysis. This method helps explain the processes and outcomes of phenomena through comprehensive observation, reconstruction, and analysis of the cases under investigation (Tellis, 1997).

The following section describes cases of three entrepreneurial ventures chosen from social, environmental, and sustainable entrepreneurship. Their entrepreneurial processes and contexts are analyzed using the appropriate theoretical frameworks to draw insights into their practices and contribution in creating social equity and CE.

10.3 SUSTAINABILITY-ORIENTED ENTREPRENEURSHIP CASES

a. **Social Entrepreneurship**

The emergence of social entrepreneurship can be attributed to growing global wealth disparity, growth of corporate social responsibility, social-market failures, technological advancements, and shared responsibility (Zahra, Rawhouser, Bhawe, Neubaum, & Hayton, 2008). Social-market failures occur when the long-standing basic needs of people, such as poverty, hunger, unclean water, unemployment, education, and human rights, are not fulfilled either by mainstream businesses or governments (Austin, Stevenson, & Wei-Skillern, 2006). Since catering to these needs is not an attractive proposition from a profitability perspective, mainstream businesses generally do not show interest in serving these markets. Governments ignore them due to a lack of public support and limited resources at their disposal. Social entrepreneurs attempt to fulfill these needs with market-based mechanisms by marshaling the necessary resources, aiming for sustainable solutions. Social entrepreneurship is the pursuit of opportunities through innovative use or a combination of resources to catalyze social change and address social needs (Mair & Marti, 2006). Hence, social entrepreneurship emphasizes creating social value instead of personal or shareholder wealth (Zadek & Thake, 1997). Depending on the nature of social problems being addressed, social entrepreneurship can occur in the nonprofit, for-profit, government sectors (Austin, Stevenson, & Wei-Skillern, 2006). Social entrepreneurs are driven by the aspirations such as wealth giving or sharing or community development, and they cater to the needs of underserved markets and socially disadvantaged populations (Martin & Osberg, 2007). So, in addition to providing goods and services, they also promote social equity and adopt innovative ways to deliver the services (Kummitha, 2018). The success of a social venture is measured in terms of social capital, social change, and fulfillment of social needs. Since social entrepreneurs use the inherently conflicting for-profit and nonprofit logic, they face institutional pressures to demonstrate legitimacy in commercial and social domains (Dacin, Dacin, & Tracey, 2011).

Frontier Markets Consulting Pvt Ltd

If the rural families, the epicenter of India's productivity and agri-revolution, do not get access to good quality products and services that deal with electricity, health care, clean water, there always will be a sense of vulnerability and we will never get to see the shift in power to rural economy. Frontier Markets was created to bridge the gap in the supply chain.

Ajaita Shah
Founder & CEO

Context – Underserved Rural Population

The majority of India's population is both rural and poor. According to world bank estimates, in 2020, nearly 65% of the population, about 897 million, live in villages. Like other emerging economies, rural India faces extreme resource scarcity. Since the rural population lacks access to electricity, clean water, infrastructure, and economic opportunities, they remain underdeveloped and underserved.

Operations

FM, established in 2011, is a social venture providing last-mile distribution solutions to rural consumers. FM uses a network of 10,000 rural women entrepreneurs (Saral Jeevan Sahelis) as the last-mile link between consumers and sellers to sell a wide range of products and services curated for rural consumers. FM's product offerings include solar energy solutions, consumer durables, smartphones, and agricultural products. Since the COVID-19 pandemic, the venture has been offering essential household items such as masks, sanitizers, and other fast-moving consumer goods, responding to the need of the customers to transact cashless and doorstep delivery. FM also offers digital services in banking and finance by partnering with Fintech companies. FM has evolved into a data-driven, women-focused product and service provider operating with a Business-to-Business-to-Consumer e-commerce platform. FM's Meri Saheli mobile app helps Sahelis to collect orders and process them. The mobile app allows FM to monitor shipment tracking, purchase orders processing, digital transactions, and customer requests on a real-time basis. In 2020, FM operated 20 warehouses, 70 delivery teams, and 20 service technicians to cover 2000 villages. FM collects quantitative and qualitative data through its digital platform, Saheli network, and call center. These data are used to obtain consumer insights about demand potential for products and services, competitive environment, and sales forecasts.

Sustainability orientation of individuals coupled with general knowledge of the natural and social environments influence them to recognize social or ecological problems (Kuckertz & Wagner, 2010; Shepherd & Patzelt, 2011). The founder had worked for 6 years in the microfinance sector, focusing on the rural population before starting FM. The job allowed her to travel to nearly 5000 villages across India and understand the rural population's needs and the challenges of fulfilling them. She had noted that although product solutions in energy and agriculture were available, they were not reaching rural households. She established FM to bridge the supply chain gap by connecting the suppliers with the rural customers. FM handled the last-mile delivery using its warehouse infrastructure, delivery, and post-sales services teams.

Her belief that the empowerment of rural women by providing them income-earning opportunities could alleviate poverty led her to collaborate with nearly 10,000 women entrepreneurs, known as "Sahelis" (helpers), in establishing sales and distribution networks. The Sahelis were identified and recruited with the help of Non-Government Organizations (NGOs) operating at the community level. FM trained them in sales, marketing, and post-sales service functions, as well as ensured they could use a mobile app to collect orders and process them. Each Saheli, on average, served 50 households and sold about 131 products or services in a year. Sahelis earned income through commission on products sold and payments by conducting surveys sponsored by large companies in rural areas. By 2020, the annual average income of a Saheli was INR 45,000, an amount that was three times higher than any income-generating activity she would have ever had. FM encouraged rural communities through training programs to use eco-friendly solar products, thus creating a positive environmental impact. The venture had sold nearly 850,000 solar solutions by 2020. In 2021, nearly 10,000 women entrepreneurs (Sahelis) used this platform to orient and connect 350,000 customers who have purchased 2.2 million solutions. FM was operating 13 branches in Rajasthan and 13 in Uttar Pradesh & Bihar, employing 145 staff members. The e-commerce platform enables FM to track orders, changing customer demands on a real-time basis.

The products and service offerings from FM have evolved to cater to the changing needs of customers. In 2011, FM sold solar energy products after observing that many rural homes were off the national electricity grid. When the electricity needs got fulfilled in 2014, the FM moved to sell durable consumer products. The government of India's demonetization in 2016 led to the increased demand for digital products and services, and FM responded by offering smartphones, internet services, and Fintech solutions by leveraging the rural coverage of 4G telecom networks and declining prices of smartphones in India. In 2018, FM introduced a mobile app – Meri Saheli app – that helped Sahelis collect data on rural households' demands & future needs, demonstrate products, and process orders & payments. The app features include local languages, voice-to-text features, and Artificial Intelligence training bots to help Sahelis and rural customers. When the COVID-19 pandemic struck in 2019, FM provided the much-needed last-mile connectivity in delivering essentials of households and grocery products. The evolutionary path of FM shows the venture's commitment to lifetime value creation for the customer. The revenue for FM is generated through sales facilitation charges, sales commission for the products sold through its platform, and marketing fees charged on product manufacturers to conduct market research.

Analysis of the Entrepreneurial Solutions

The extent to which the society perceives that the existing institutional setup could not fulfill specific social needs, any entrepreneurial solutions to address those needs are valued (Mair & Marti, 2006). FM's mission is to facilitate rural customers to access the products and services at affordable prices without compromising quality. Since the marketplace is scattered and the transactions occur in a subsistence environment, most product manufacturers and service providers found it challenging to identify, influence, and serve consumer needs, leading to a gap in the supply chain. The business goal of FM was to bridge the supply chain gap so that it would be beneficial to both the consumers and the suppliers. However, FM had to address the challenges associated with rural markets, such as a non-munificent environment and other socio-economic issues (Mair & Marti, 2006).

FM operated in resource-constrained environments and addressed the associated challenges by employing bricolage mechanisms (Bojica, Ruiz Jiménez, Ruiz Nava, & Fuentes-Fuentes, 2018). Bricolage is "making do by applying combinations of the resources at hand to new problems and opportunities" (Baker & Nelson, 2005, p. 333). Bricolage disregards common limitations associated with the available resources but tries solutions using them and dealing with the results. Social entrepreneurs use any such resources that can be obtained for free or at a minimal cost. FM effectively used the infrastructure already created by government agencies, NGOs, commercial organizations, and Self Help Groups (SHGs) in the rural markets. For example, the Government of

India's National Rural Livelihood Mission (NRLM) led to the growth of SHGs federated at village and cluster levels. The NRLM helped the rural poor in capacity-building activities and provided them access to institutional credit. Many NGOs have long worked at the community level in the rural sector to improve the socio-economic status of the rural population. Microfinance institutions and other commercial organizations have developed distribution and service networks to cater to the needs of rural markets. FM valued the already-created ecosystem, and it approached the agencies with a business proposition embedded with social value. Since the agencies found value alignment with FM's proposition, they agreed to collaborate with FM. For example, NGOs linked with SHGs helped FM identify women with social capital and the potential to become entrepreneurs (Sahelis). Since the literacy levels of poor rural consumers are low, business exchanges need to be fair and transparent in these markets, FM leveraged the social capital of the Sahelis to develop its customer proposition revolving around lifetime value creation (Viswanathan, Gajendiran, & Venkatesan, 2008).

Bricolage is to combine and apply the resources for new purposes. When the COVID-19 pandemic struck, FM, with the permission of government agencies, repurposed its distribution network to deliver essential items and provide digital services at the doorsteps of rural households. Although the distribution network was not originally intended to deliver essential items, combining resources for new and different applications can lead the social entrepreneurs to innovate service deliveries. Social ventures, unlike commercial ventures, found more scope and opportunities to create a social impact during crisis such as COVID-19.

b. **Eco-Entrepreneurship**

Human societies and the businesses operating within them consume finite natural resources. Entrepreneurial actions lead to innovation and economic growth, but they also unintentionally degrade natural resources. However, entrepreneurs identify and pursue the value-creating opportunities that degrading natural resources offer so that environmental damage mitigation or even reversal of those damages can occur. The objective of eco-entrepreneurship is to pursue technological and nontechnological innovations to reduce, recycle, and reuse natural resources, spread environmental awareness, and create a greener, more eco-friendly economy. The emerging areas for eco-entrepreneurship include renewable energy, energy-efficient buildings, clean water, depletion of natural resources, air pollution, loss of biodiversity, automobile emissions, and climate change. Eco-preneurial ventures are designed to be environmentally responsible from scratch in their process, product, and service offerings (Isaak, 2017). The development of eco-entrepreneurship depends on stricter environmental standards & regulations, demand for eco-friendly products & services, investors' interest to invest in eco-conscious ventures, institutional & policy support in developing suitable technologies, the strength of green activism, and environmental jurisprudence. Ecopreneurs are committed to reduce or reverse environmental damage through business solutions. Eco-entrepreneurship facilitates communities to move toward sustainability by adopting corrective measures to address ecological opportunities and challenges.

Doodlage

The problem in the fashion industry starts from the fact that it is based on the linear fashion model…Its about how we want to do it.

Kriti Tula
Creative Director & Co-founder
Doodlage

Context – Environmental Cost of the Linear Model of Textile and Clothing Industry

The T&C industry operations have been scrutinized globally for their disproportionate use of precious water supply and the environmental impact. Each stage of the industry's supply chain has environmental implications for water, chemical, energy usage, CO_2 emissions, and solid waste generations. The T&C industry is one of the major consumers of fresh water in the world. The water required to produce 1 kg of cotton is about 20,000 L, and processing one ton of dyed fabric can take up to 200 tons of freshwater. The fertilizer and pesticides used in growing cotton crops and the untreated toxic wastewater released by the clothing factories pollute water resources. Nevertheless, the industry continued to grow, partly due to the rise of fast fashion, a trend characterized by falling clothing prices, higher speed of fashion delivery, and short-lived garment use (Niinimäki, Peters, Dahlbo, Perry, Rissanen, & Gwilt, 2020).

Clothing companies pursued aggressive pricing strategies and streamlined their operations to decrease clothing prices relative to other consumer goods. Shorter production lead times allow the clothing brands to introduce new lines more frequently. The annual garment production rate was doubled between 2000 and 2014 (Remy, Speelman, & Swartz, 2016). The greater variety offered at lower prices induced consumers to buy more than they needed and waste more. Less than half of used clothes are collected for reuse or recycling. Only 1% are recycled into new clothes resulting in large amounts of textile waste, most of which are either incinerated or landfilled. Nearly 73% of preproduction textile waste gets landfilled due to the absence of sustainable methods to reuse the fabric waste. Since India is one of the largest producers of fabric and garments, discarded waste has a substantial environmental impact. The T&C industry is the largest source of waste after paper, plastic, and compost.

The T&C waste can be, depending upon the type of sources, broadly classified into three main types – (i) postindustrial waste – waste generated from cloth and garment manufacturing process, (ii) preconsumer waste – defective garments rejected by manufacturer or distributor, unsold garments at the retail stores, and (iii) postconsumer waste – worn out, unwanted, and damaged clothes discarded by consumers. The key challenge in reducing the amount of textile waste lies in inducing the stakeholders of the T&C industry to adopt the CE principles – 3R.

Operations

After observing the considerable amount of fabric waste generated from garment factories, the founder believed that the preconsumer textile waste could be sustainably converted into fashion wear. Her commitment to reducing environmental pollution due to the T&C industry led to establishing Doodlage in 2012. The venture uses the reuse principle of CE to upcycle postindustrial waste into limited editions of garment collections using innovative designs. Doodlage creates a trendy fashion wear-out of surplus fabric from manufacturers, who usually scrap about 11% of the fabric due to inefficient cutting patterns. These short but well-finished editions of the high-end garment collection save tons of fabric waste from being landfilled.

Doodlage buys pieces of fabric of any size under 100 meters discarded by the garment factories. The fabric waste is being sold based on its weight and costs one-third of the price of new fabric. However, the money saved in the purchase of fabric waste is used to fix the fabric by Doodlage. The production of a premium range of garments out of this fabric involves creative use of mixing, patching, embroidering, and handweaving by the team of artisans. The team of artisans employed by the venture first works on the garment designs to bring the fabric pieces together and fix the garments' pieces.

The venture achieves zero waste by using the wastage generated from its production line in every possible method. The fabric waste undergoes three cycles of the production process, and the value is added to the waste generated in every cycle. After making the garments from the preproduction fabric waste, the leftover is paneled together to get more fabric yardage. Depending on the durable strength, the paneled fabric is then converted into home furnishings, bags, cushion covers, table runners, or paper-based products such as garment hangtags and stationery. The hangtags are used to pack repair kits with which the customers can fix their garments to last longer. The remaining waste is used to create packaging item tote bags that the customers can reuse until they last, thus achieving zero waste in the whole operation. Apart from fabrics, Doodlage sources zippers in bulk quantity from the trims that factories discard as surplus. Doodlage offers its collection for both women and men, and some of the collections are gender-neutral. In 2018, Doodlage had access to 40 multidesigner stores in India and a few other international stores. The venture sells garments through its website and other e-commerce platforms. The venture collaborates with garment retail stores to offer solutions to manage consumer returns and store defectives by repairing upcycling processes. For example, Decathlon's Bengaluru store collaborates with Doodlage to create products that have been upcycled or repaired using garment waste and customer returns from the store. Under the "Repaired products" category, the store resells the upcycled products, including t-shirts, accessories, swimwear, and sports goods.

Analysis of Ecopreneurial Solutions

Although entrepreneurship induces innovations and generates economic growth, it also causes the unintended consequence of environmental degradation. The environmental degradation due to the T&C industry is the unintended consequence of previous economic innovations in industrialized production methods employed in the agricultural and T&C sectors. However, these methods are notorious for exploiting precious ecological resources and polluting them. The fast-fashion trend thriving in the socio-cultural system incentivizes the economic agents to operate on a linear model of "take, make, and waste," unmindful of its implications for environmental sustainability. However, entrepreneurs are also quick to respond to the value-creating opportunities arising from environmental degradation (Boons & Wagner, 2009). Entrepreneurs create and use new markets to facilitate those affected to buy goods and services that help us fix the environmental problems partly and enough to avoid disaster (Potts, Foster, & Straton, 2010).

The role of entrepreneurship in addressing environmental sustainability issues can be analyzed using the micro, meso, and macro frameworks employed in the coevolutionary model. (Dopfer, Foster, & Potts, 2004; Dopfer, 2005; Dopfer & Potts, 2009). According to this model, the economic, political, socio-cultural, creative, and ecological systems are interconnected, interact, and coevolve to produce an economic order. Economic evolution occurs when the novel entrepreneurial actions applied at the microeconomic level result in new rules of value creation. Since economic activities are embedded in an ecological context, any change in them affects the ecological system. The economic system, driven by entrepreneurial activities, continuously evolves to lower environmental impact. However, widespread adoption of such evolutionary changes resulting in mitigation or reversal of environmental degradation depends upon the other supporting changes in political, socio-cultural, and creative systems. For example, when green activists identify and publicize the environmental impacts of economic activities, it may lead to widespread awareness about the environmental issues, new thoughts, and actions to address them. Upon observing the support for such thoughts and actions in the society, the political franchisees change the "rules of the game" through fiscal and regulatory measures, such as implementing stricter environmental laws, treaties, and agreements. The creative system comprising scientists, writers, inventors, artists, and philosophers to invent a new combination of machine, material, and organizational resources to reduce environmental degradation. The system of values, beliefs, traditions, trust, and social climate shared by the majority of the members in the society shapes the socio-cultural system. Hence, environmental awareness and a positive attitude toward environmental protection may lead consumers to opt for eco-friendly products and services. Thus, the socio-cultural, political, and creative changes induce entrepreneurs in the economic system to offer product and service solutions in the market to reduce environmental impact. For example, stricter environmental regulations in the political system can lead to technical innovations and entrepreneurship (Rennings, 2000; Beise & Rennings, 2005). E-waste regulations introduced in 2016 in India stimulated the growth of producer responsible organizations who offer extended producer responsibility services to global brands. Such entrepreneurial solutions lead to more responsible e-waste management practices, reducing environmental degradation. The vehicle scrapping policy introduced in 2021 provides entrepreneurial opportunities in creating the necessary infrastructure in the form of automated testing stations and vehicle scrapping facilities.

The mechanisms used in these coevolving, interconnected systems need to reinforce and align to achieve environmental sustainability. Doodlage's operations positively influence economic, creative, and socio-cultural systems to move toward environmental sustainability. The venture identified a sustainable way to upcycle preproduction waste and create a new business line out of it. The production of garments involves creative work of mixing, patching, embroidering, and hand-weaving by a team of artisans, thus earning an artistic reputation for their products. The brand offerings appeal to eco-conscious consumers who may switch over to wearing eco-conscious fashion products, thus setting a new trend in motion that can positively affect socio-cultural beliefs and values toward wearing eco-conscious garments. The association with Decathlon's Bengaluru store shows the adoption of entrepreneurial innovation pioneered by Doodlage by other players operating in the preconsumer waste segment.

c. **Sustainable Entrepreneurship**

Sustainable entrepreneurship considers the long-term environmental and social outcomes and impacts equally as essential as the short-term financial outcomes. Sustainable entrepreneurs recognize, develop, and exploit business opportunities arising out of the sustainability needs of society (Schaltegger & Wagner, 2011). Sustainable entrepreneurs aim to preserve nature, life support, and community while pursuing entrepreneurial opportunities to create economic and non-economic benefits for individuals and society. (Shepherd & Patzelt, 2011). Sustainable entrepreneurship is different from social entrepreneurship and eco-entrepreneurship in its focus and value-creation domain. Social entrepreneurship addresses unfulfilled social needs such as access to basic human needs, improved quality of life, and social equity. Eco-entrepreneurship focuses on developing a business model that does not deplete natural resources but replenishes them.

In contrast, sustainable entrepreneurship recognizes the interdependence of social, environmental, and economic domains and pursues opportunities identified at the intersection of the three. Sustainable entrepreneurs integrate social equity and environmentalism and develop financially sustainable business models. They implement strategies based upon a triple bottom line (3BL) of economics (profit), environmental (planet), and social wellbeing (people) to remove or reduce unsustainable practices in society (Dixon & Clifford, 2007). Sustainable entrepreneurs need to integrate and balance all three inherently conflicting goals. Pursuing the social or environmental goal at the cost of the economic goal can make the venture unviable. Focusing on economic goals at the cost of other goals can lead to 'mission dilution,' resulting in the loss of social and environmental values. Sustainable ventures' economic, environmental, and social goals reinforce each other and mutually define the demands. Sustainability goals are achieved by seeking synergistic opportunities at the intersection of 3BL goals. Creating social and environmental values may be closely linked or even integral to achieving financial goals, and entrepreneurial opportunities can be identified at the intersection of 3BL goals. (Kumar, 2020)

Elrhino Eco Industries Private Limited

It has always been a priority for us to remember why we are running this venture. Our commitment is towards the community, not to the business. However, the business needs to be financially sustainable also. So every day, there is a fight between commerce and cause. As sustainable entrepreneurs, we need to balance economic, social, and environmental goals.

Mahesh Chandra Bora & Nisha Bora
Founders, Elrhino

Context – Wildlife–Human Conflict & Conservation of Indian Rhino and Elephants

The great one-horned rhinoceros, popularly referred to as "Indian rhino" (rhino), is the largest rhino species found in alluvial flood plains and adjacent swamps, forests located along the Himalayas foothills. Once widely distributed in the vast Gangetic plains, their habitat is now restricted to small areas in Indo-Nepal terrain and the national parks located in West Bengal and Assam, the northeastern states of India. It is estimated that nearly 85% of the global Indian rhino population is concentrated in Assam, where Kaziranga National Park alone accounts for about 70% of the rhino population. According to the World Wide Fund for Nature – India, the conservation measures introduced in the early 1900s have helped the rhino population grow from a mere 75 individuals in 1905 to 2700 in 2012.

The state of Assam has the second-highest population of elephants in India. Elephants feed mainly on grasses, tree bark, and leaves and produce about 220 pounds of dung after spending 19 hours a day feeding. The survival of rhinos and elephants is under constant threat due to illegal poaching, habitat encroachment, WHC, and flooding due to climate, land-use changes. The looming threats prompted the International Union for Conservation of Nature to declare rhinos' vulnerable species in 2008.

Natural habitats for wildlife have decreased globally due to land-use changes initiated to meet the growing population needs. The natural habitats of rhinos and elephants were converted into agricultural fields to meet the needs of the rising human population (Dinerstein, 2003). Humans and wild animals increasingly compete over space for living and food (Lamarque, Anderson, Fergusson, Lagrange, Osei-Owusu, & Bakker, 2009; Madden, 2004). The shrinking habitat forces animals to frequently move out of protected areas' boundaries to search for food in the surrounding villages, triggering WHC.

WHC occurs when animals threaten the livelihood and safety of humans recurrently. Humans lose their crops, livestock, properties, and in some cases, even their lives due to WHC. They often kill the suspected animals and alter the wildlife habitats in response to prevent losses (Madhusudan & Karanth, 2002). Elephants and rhinos often venture out of their shrinking habitat in search of food, causing massive losses to small farmers whose fields are located around the preservation areas. Animal raids in the cultivated areas invariably lead to WHC, and the animals are often killed in retaliation.

Moreover, the losses due to animal raids increasingly make the community members living in the vicinity increasingly show an indifferent attitude toward poachers, thus leaving the animals vulnerable to the illegal activities of poachers. Since wildlife protection is valued in society, lethal retaliation against wildlife is illegal or simply not socially acceptable (Naughton-Treves & Treves, 2005). However, strict regulations related to protected areas, such as denying access to natural resources, have negative implications for the socio-economic conditions of the people (Budhathoki, 2003). Thus, effective management of WHC is crucial for the continued survival of both wildlife and local communities.

Illegal poaching and wildlife smuggling are other challenges facing endangered and vulnerable wildlife. The misperceived health and vitality effects or status effects lead to the demand for wildlife, which encourages illegal trade (Lyons & Natusch, 2013). Though there is no scientific evidence on the medicinal value of rhino horns, they are used in traditional Asian medicines for treating a range of ailments, including epilepsy, fever, and cancer. The illegal trade of rhino horns, fueled by market demand, encourages poaching and remains a significant threat to the species survival. Elephants are poached for ivory tusks. Since male Asian elephants have tusks, poaching them leads to a skewed gender ratio affecting the breeding rates of this animal. Although the ivory trade is banned globally, illegal trade is thriving due to high demand.

Operations

Elrhino was established in 2014 to spread awareness about the conservation of elephant and rhino population and the importance of inclusion of the marginalized communities who live in the vicinity while drawing out any plan to conserve any ecosystem. Elrhino operations are located in Chhaygaon, Kamrup District, Assam, and the nearby areas of Pobitora wildlife sanctuary, which has the second-largest rhino population globally. The venture engages the community members living in the vicinity of the Chhaygaon to produce handmade paper using elephant, rhino dung, and other naturally occurring fibrous waste. The community members living in the vicinity of Pobitora wildlife sanctuary are employed in raw material sourcing, transportation, manufacturing, designing, and producing a range of value-added products out of paper.

Rhinos and elephants primarily feed on long grasses and leaves. Since they have weak digestive systems, their dung is rich in fibrous material and used as raw material for making paper. The dung is washed first to remove any fecal matter and obtain the fiber. The fiber obtained from dung is boiled with other natural fibers such as jute, banana barks, and cotton hosiery rags to prepare pulp. The pulp is spread to form sheets, pressed together, dried, and then cut into standard sizes. The cut pieces are dyed to produce the final product – the food-grade paper free from any wood or chemicals. Elrhino produces a range of aesthetically appealing products, including notebooks, journals, playing cards, bags out of this eco-friendly paper. The whole manufacturing operation is carried out on zero-waste principles. The manufacturing unit located in Chhaygaon village is operating with 15 full-time employees. Depending on the demand, the venture outsources specialized work activities such as sourcing fibrous raw material and developing indigenous designs to 35 more artisans in the community.

Analysis of Entrepreneurship Solutions

The management of natural resources essentially focuses on managing people whose livelihoods are dependent on the resources. However, the importance of the social aspects of natural resource management is not widely recognized. Conservation efforts without local community welfare considerations can become ineffective, leading to reduced cooperation and continued conflicts. The involvement of local communities in resolving WHC may lead to benefits for both the locals and the wildlife. The goal is to facilitate the coexistence of wildlife and humans by balancing the conservation goals and the community's needs who live in the nearby areas.

Elrhino implements strategies based upon a 3BL of economics (profit), environmental (planet), and social wellbeing (people) goals. Social goals are focused on improving the livelihoods of communities living in the vicinity of the habitat of the animals by engaging them in the manufacturing of environment-friendly paper-based products. The environmental goals focus on protecting elephants and rhinos along with forest resources. Since rhinos and elephants share their habitat with other flora and fauna, any effort to conserve them protects the associated ecosystem. Elrhino's business model is developed based on the assumption that the engagement of stakeholders – the local community members in the conservation of wildlife and forest resources can lead to avoidance of WHC.

WHC needs to be addressed holistically, considering the underlying social, cultural, and economic contexts. The habitats of rhinos declared as protected areas in India are surrounded by dense human settlements. Hence, it is critical to ensure that the communities living around rhino reserves develop nurturing attitudes toward rhinos. The livelihoods of communities can be sustained when they benefit from the revenue generated through eco-friendly production activities. Elrhino makes the local community members stakeholders of animals' wellbeing by engaging them in its operations. Most of the work is done by them at homes and complements their traditional practices. The community members are engaged through the commercial operations of paper making out of naturally occurring raw material in the forest, which provides them the much-needed income that helps to achieve sustainable livelihoods. Since their livelihoods are now dependent on natural capital, they no longer compete with animals for natural resources but learn to coexist with wildlife. They develop a positive attitude toward protecting wildlife and forest resources, leading to the avoidance of WHC.

India has such communities that would fiercely guard the wildlife and natural resources present in their locality. For example, the Bishnois community in Rajasthan, a western state of India, is widely known for its strong commitment to environmental values, which manifests in strict adherence to the complete ban on killing animals and cutting down trees. As rhinos are revered animals and a symbol of pride in Assam, conservation efforts often inspire environmental stewardship among the communities who co-habitat with the animal. Elrhino's operations provide a sense of purpose and instill pride in protecting the natural capital. Although the community members feel the need to preserve the ecosystem, Elrhino removes any barriers so that the symbiotic relationship between humans and ecosystems flourish further. The inclusive nature of the communities whose lives depend on the ecosystem is acknowledged and addressed in the solutions. The community members get an identity and voice through Elrhino's operations. The relevant stakeholders, including government agencies, customers, and society, take cognizance of their needs and interests. Elrhino's products communicate the possibility of balancing the needs of both wildlife and humans through sustainable methods. They gain the attention of the relevant stakeholders both locally and globally to the social needs of marginalized communities living in the vicinity of eco-conservation projects.

10.4 IMPLICATIONS FOR RESEARCH AND PRACTICE

The current study has implications for both research and practice in the domain of sustainability-oriented entrepreneurship. The process-tracing methods of a case study research shall be used to understand the impacts of green activists on both the regulating agencies and legislatures to implement new policies and rules to protect the environment. The future research shall study the cases of those entrepreneurs who identify and pursue the market opportunities that degrading natural resources offer. Future researchers can collect more evidence to examine whether creating social and environmental values are closely linked or even integral to achieving financial goals.

Sustainability-related challenges offer many market opportunities for entrepreneurs to pursue profitably. Entrepreneurs may use bricolage mechanisms and leverage social capital and technology to create value for the target customers in resource-constrained environments. Entrepreneurs shall use the CE principles of 3R to create new value to the waste material through innovative production methods. Any effort to conserve natural resources without the local community welfare can become ineffective and may lead to reduced cooperation and continued WHCs. The engagement of local communities in the pursuit of sustainable entrepreneurship can reduce conflicts and benefit both the locals and the wildlife. Such entrepreneurial ventures shall focus on the opportunities that can be found at the intersection of wildlife protection, natural resource conservation, and the competencies of local communities.

10.5 CONCLUSION

Sustainability challenges offer new opportunities for creative human actions. Sustainability is not just restricted to being environment friendly, but also about wealth creation and job creation. The analysis of entrepreneurial solutions offered by the ventures – FM, Doodlage, and Elrhino show that ecological protection, economic development, and social equity are inextricably linked. Financial wealth does not lead to sustainability, whereas poverty is often the root cause of unsustainability. Hence, it is essential to integrate social equity into the practice of ecological sustainability (Hamstead & Quinn, 2005). The case of Elrhino suggests that natural capital conservation and community development may be critical in strengthening the local economy. The central notion of the ecological-economic framework – healthy ecosystems are critical to the long-term financial wellbeing of communities is demonstrated in the practices of Elrhino (Roseland, 2000). Eco-entrepreneurship kindles environmental stewardship among the community members whose livelihoods are dependent on natural resources. Environmental stewardship reflects ethical responsibility that incorporates helping behaviors triggered by individual's awareness of the interconnectedness between their behavior, other behavior, and environmental effects (Worrell & Appleby, 2000).

The case of Doodlage shows how the CE principles of 3R can offer potential opportunities for eco-entrepreneurship. Doodlage uses upcycling method to add new value to the waste material through innovative production methods. Market forces alone cannot affect the transition from a linear economy to a CE. Sustainability-oriented entrepreneurs play a significant role in changing society's normative and behavioral aspects toward a CE. They create a new economic order using innovative products and services that acknowledge ecological limitations, bring social equity, and balance economic, environmental, & social goals.

Most often, social entrepreneurs operate in resource-constrained situations, where mobilizing resources and achieving legitimacy become challenging. However, as demonstrated by FM, entrepreneurs use bricolage mechanisms to marshal the necessary resources to create social value. The entrepreneurial solutions offered by sustainable ventures discussed in this chapter indicate a range of activities used to create social and environmental value beyond economic values. The activities include bricolage mechanisms, education, awareness, training, capacity building, and leveraging technology. The government and its institutions have well supported the ventures.

Sustainable entrepreneurs often create niche markets by differentiating their products based on social and environmental characteristics. They position their products in the premium segment, emphasize quality over price, save costs by removing intermediaries along the supply chain, and use e-commerce channels (Belz & Binder, 2017; Choi & Gray, 2008). The products of Elrhino and Doodlage ventures are positioned in a niche segment and differentiated based on social and environmental characteristics. Sustainable entrepreneurs gain legitimacy using their sustainability mission and leverage it while managing with internal and external stakeholders. The sustainability mission helps them marshal the necessary resources, network with the stakeholders, create demand for their products & services, and influence the policymakers to create supporting infrastructure.

REFERENCES

Andriantiatsaholiniaina, L. A., Kouikoglou, V. S., & Phillis, Y. A. (2004). Evaluating strategies for sustainable development: Fuzzy logic reasoning and sensitivity analysis. *Ecological Economics*, 48(2), 149–172. https://doi.org/10.1016/j.ecolecon.2003.08.009.

Austin, J., Stevenson, H., & Wei-Skillern, J. (2006). Social and commercial entrepreneurship: Same, different, or both?. *Entrepreneurship Theory and Practice*, 30(1), 1–22. https://doi.org/10.1111/j.1540-6520.2006.00107.x.

Baker, T., & Nelson, R. E. (2005). Creating something from nothing: Resource construction through entrepreneurial bricolage. *Administrative Science Quarterly*, 50(3), 329–366. https://doi.org/10.2189/asqu.2005.50.3.329.

Beise, M., & Rennings, K. (2005). Lead markets and regulation: A framework for analyzing the international diffusion of environmental innovations. *Ecological Economics*, 52(1), 5–17. https://doi.org/10.1016/j.ecolecon.2004.06.007.

Belz, F. M., & Binder, J. K. (2017). Sustainable entrepreneurship: A convergent process model. *Business Strategy and the Environment*, 26(1), 1–17. https://doi.org/10.1002/bse.1887.

Bojica, A. M., Ruiz Jiménez, J. M., Ruiz Nava, J. A., & Fuentes-Fuentes, M. M. (2018). Bricolage and growth in social entrepreneurship organisations. *Entrepreneurship & Regional Development*, 30(3–4), 362–389. https://doi.org/10.1080/08985626.2017.1413768.

Boons, F., & Wagner, M. (2009). Assessing the relationship between economic and ecological performance: Distinguishing system levels and the role of innovation. *Ecological Economics*, 68(7), 1908–1914. https://doi.org/10.1016/j.ecolecon.2009.02.012.

Budhathoki, P. (2003). A category V protected landscape approach to buffer zone management in Nepal. *Parks*, 13(2), 22–30. http://citeseerx.ist.psu.edu/viewdoc/download?doi=10.1.1.475.2738&rep=rep1&type=pdf#page=24.

Choi, D. Y., & Gray, E. R. (2008). The venture development processes of "sustainable" entrepreneurs. *Management Research News*, 31(8). https://www.emerald.com/insight/content/doi/10.1108/01409170810892127/full/pdf?title=the-venture-development-processes-of-sustainable-entrepreneurs.

Cohen, B., & Winn, M. I. (2007). Market imperfections, opportunity and sustainable entrepreneurship. *Journal of Business Venturing*, 22(1), 29–49. https://doi.org/10.1016/j.jbusvent.2004.12.001.

Dacin, M. T., Dacin, P. A., & Tracey, P. (2011). Social entrepreneurship: A critique and future directions. *Organization Science*, 22(5), 1203–1213. https://doi.org/10.1287/orsc.1100.0620

Dinerstein, E. (2003). *The Return of the Unicorns*. Columbia University Press.

Dixon, S. E. A. and Clifford, A. (2007). Ecopreneurship – a new approach to managing the triple bottom line, *Journal of Organizational Change Management*, 20(3), 326–345. https://doi.org/10.1108/09534810710740164.

Dopfer, K. (2005). Evolutionary economics: A theoretical framework. *The Evolutionary Foundations of Economics*, 3–55. https://www.researchgate.net/profile/Kurt-Dopfer/publication/297557333_The_Evolutionary_Foundations_of_Economics/links/57b42ce408ae19a365fadb3c/The-Evolutionary-Foundations-of-Economics.pdf.

Dopfer, K., Foster, J., & Potts, J. (2004). Micro-meso-macro. *Journal of Evolutionary Economics*, 14(3), 263–279. https://doi.org/10.1007/s00191-004-0193-0.

Dopfer, K., & Potts, J. (2009). On the theory of economic evolution. Evolutionary and institutional *Economics Review*, 6(1), 23–44. https://doi.org/10.14441/eier.6.23.

Hamstead, M. P., & Quinn, M. S. (2005). Sustainable community development and ecological economics: Theoretical convergence and practical implications. *Local Environment*, 10(2), 141–158. https://doi.org/10.1080/1354983052000330743.

Isaak, R. (2017). *Green Logic: Ecopreneurship, Theory and Ethics*. Routledge.

Kuckertz, A., & Wagner, M. (2010). The influence of sustainability orientation on entrepreneurial intentions – Investigating the role of business experience. *Journal of Business Venturing*, 25(5), 524–539. https://doi.org/10.1016/j.jbusvent.2009.09.001.

Kumar, S. S. (2020). Managing paradoxical tensions at the intersection of the triple bottom line goals: The case of under the mango tree. *International Review of Entrepreneurship*, 18(4).

Kummitha, R. K. R. (2018). Social entrepreneurship, energy and urban innovations. *Mainstreaming Climate Co-Benefits in Indian Cities*, Springer, Singapore, pp. 265–283. https://doi.org/10.1007/978-981-10-5816-5_11.

Lamarque, F., Anderson, J., Fergusson, R., Lagrange, M., Osei-Owusu, Y., & Bakker, L. (2009). *Human-Wildlife Conflict in Africa: Causes, Consequences and Management Strategies*. Food and Agriculture Organization of the United Nations (FAO), 157.

Lyons, J. A., & Natusch, D. J. (2013). Effects of consumer preferences for rarity on the harvest of wild populations within a species. *Ecological Economics*, 93, 278–283. https://doi.org/10.1016/j.ecolecon.2013.06.004.

Madden, F. (2004). Creating coexistence between humans and wildlife: Global perspectives on local efforts to address human–wildlife conflict. *Human Dimensions of Wildlife*, 9(4), 247–257. https://doi.org/10.1080/10871200490505675.h

Madhusudan, M. D., & Karanth, K. U. (2002). Local hunting and the conservation of large mammals in India. *Ambio*, 49–54. https://www.jstor.org/stable/4315210.

Mair, J., & Marti, I. (2006). Social entrepreneurship research: A source of explanation, prediction, and delight. *Journal of World Business*, 41(1), 36–44. https://doi.org/10.1016/j.jwb.2005.09.002

Martin, R. L., & Osberg, S. (2007). Social entrepreneurship: The case for definition. *Standford Social Innovation Review*, 28–39. https://www.ngobiz.org/picture/File/Social%20Enterpeuneur-The%20Case%20of%20Definition.pdf.

McDonough, W., & Braungart, M. (2002). Design for the triple top line: New tools for sustainable commerce. *Corporate Environmental Strategy*, 9(3), 251–258. https://doi.org/10.1016/S1066-7938(02)00069-6.

Moreno, M., Braithwaite, N., & Cooper, T. (2014). Moving beyond the circular economy. *Proceedings of Going Green-CARE Innovation*, 1–10.

Naughton-Treves, L., & Treves, A. (2005). Socio-ecological factors shaping local support for wildlife: Crop-raiding by elephants and other wildlife in Africa. *Conservation Biology Series-Cambridge*, 9, 252. https://faculty.nelson.wisc.edu/treves/pubs/2005NaughtonTrevesLTrevesA.pdf.

Niinimäki, K., Peters, G., Dahlbo, H., Perry, P., Rissanen, T., & Gwilt, A. (2020). The environmental price of fast fashion. *Nature Reviews Earth & Environment*, 1(4), 189–200. https://doi.org/10.1038/s43017-020-0039-9.

Park, J., Sarkis, J., & Wu, Z. (2010). Creating integrated business and environmental value within the context of China's circular economy and ecological modernization. *Journal of Cleaner Production*, 18(15), 1494–1501. https://doi.org/10.1016/j.jclepro.2010.06.001.

Potts, J., Foster, J., & Straton, A. (2010). An entrepreneurial model of economic and environmental co-evolution. *Ecological Economics*, 70(2), 375–383. https://doi.org/10.1016/j.ecolecon.2010.09.009.

Remy, N., Speelman, E., & Swartz, S. (2016). *Style that's Sustainable: A New Fast-Fashion Formula*. McKinsey Global Institute. https://www.mckinsey.com/business-functions/sustainability/our-insights/style-thats-sustainable-a-new-fast-fashion-formula.

Rennings, K. (2000). Redefining innovation – eco-innovation research and the contribution from ecological economics. *Ecological Economics*, 32(2), 319–332. https://doi.org/10.1016/S0921-8009(99)00112-3.

Roseland, M. (2000). Sustainable community development: Integrating environmental, economic, and social objectives. *Progress in Planning*, 54(2), 73–132. https://doi.org/10.1016/S0305-9006(00)00003-9.

Schaltegger, S., & Wagner, M. (2011). Sustainable entrepreneurship and sustainability innovation: Categories and interactions. *Business Strategy and the Environment*, 20(4), 222–237. https://doi.org/10.1002/bse.682.

Schumpeter, J. (1942). Creative destruction. *Capitalism, Socialism and Democracy*, 825, 82–85.

Shepherd, D. A., & Patzelt, H. (2011). The new field of sustainable entrepreneurship: Studying entrepreneurial action linking "what is to be sustained" with "what is to be developed". *Entrepreneurship Theory and Practice*, 35(1), 137–163. https://doi.org/10.1111/j.1540–6520.2010.00426.x.

Tellis, W. (1997). Application of a case study methodology. *The Qualitative Report*, 3(3), 1–19.

Viswanathan, M., Gajendiran, S., & Venkatesan, R. (2008). *Enabling Consumer and Entrepreneurial Literacy in Subsistence Marketplaces*, 12. Springer Science & Business Media.

WCED, S. W. S. (1987). World commission on environment and development. *Our Common Future*, 17(1), 1–91.

Worrell, R., & Appleby, M. C. (2000). Stewardship of natural resources: Definition, ethical and practical aspects. *Journal of Agricultural and Environmental Ethics*, 12(3), 263–277. https://doi.org/10.1023/A:1009534214698.

Yin, R. (1984). *Case Study Research*. Beverly Hills.

Zadek, S., & Thake, S. (1997). *Practical People, Noble Causes. How to Support Community-Based Social Entrepreneurs*. New Economics Foundation, London.

Zahra, S. A., Rawhouser, H. N., Bhawe, N., Neubaum, D. O., & Hayton, J. C. (2008). Globalization of social entrepreneurship opportunities. *Strategic Entrepreneurship Journal*, 2(2), 117–131. https://doi.org/10.1002/sej.43.

Index